The Geology

of the Parks

Monuments

and Wildlands

of Southern Utah

Including ≈

the Henry Mountains

Grand Staircase–Escalante National Monument

Capitol Reef National Park

Bryce Canyon National Park

and Zion National Park

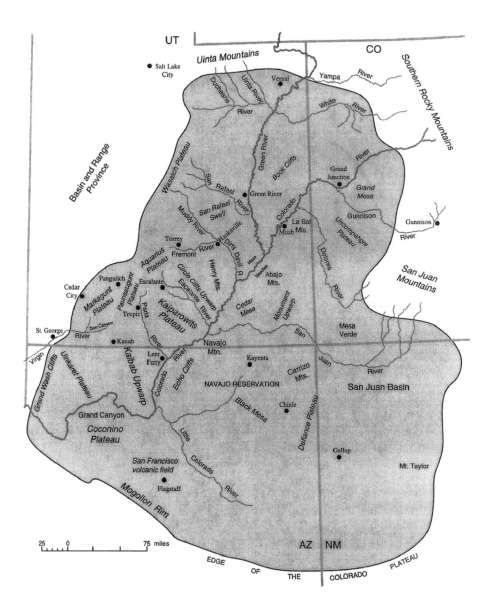

UT

CO

Salt Lake
City

Uinta Mountains

Southern Rocky Mountains

Duchesne

Uinta River

Vernal

Yampa

River

River

White

River

Wasatch Plateau

Green River

Book Cliffs

Grand
Junction

River

San Rafael River

Green River

Grand
Mesa

Muddy River

San Rafael
Swell

River

Colorado

La Sal
Moab Mts.

Gunnison

Gunnison

River

Basin and Range
Province

Hanksville

Torrey

Aquarius
Plateau

Fremont

River

Dirty Devil R.

Wash

Uncompahgre
Plateau

San Juan
Mountains

Cedar
City

Panguitch

Paunsaugunt Plateau

Escalante

Circle Cliffs Upwarp

Escalante River

Henry Mts.

Needles

Abajo
Mts.

Dolores

River

Makagunt
Plateau

Tropic

Paria

Kaiparowits
Plateau

Cedar
Mesa

Monument
Upwarp

Mesa
Verde

St. George

Zion Canyon

River

Kanab

River

Navajo
Mtn.

San

Virgin

Lees
Ferry

River

Echo Cliffs

Kayenta

Juan

River

Grand Wash Cliffs

Uinkaret Plateau

Kaibab Upwarp

Colorado

NAVAJO RESERVATION

Carrizo
Mts.

San Juan Basin

Grand Canyon

Black Mesa

Chinle

Defiance Plateau

Coconino
Plateau

Little

San Francisco
volcanic field

Colorado

Gallup

Mt. Taylor

Flagstaff

River

Mogollon Rim

25   0   75 miles

AZ   NM

EDGE   OF   THE   COLORADO   PLATEAU

The Colorado Plateau

The Geology
of the Parks
Monuments
and Wildlands
of Southern Utah

Robert Fillmore

≈

The University of Utah Press

Salt Lake City

2000 01 02 03 04 05 06

5 4 3 2 1

LIBRARY OF CONGRESS CATALOGING-IN-PUBLICATION DATA

Fillmore, Robert, 1957–
    The parks, monuments, and wildlands of southern Utah : a
geologic history with road logs of highways and major backroads
/ Robert Fillmore.
        p. cm.
    Includes bibliographical references (p.)
    ISBN 0-87480-652-6
        1. Geology—Utah—History. 2. Parks—Utah. 3. Monuments—
Utah. 4. Wilderness areas—Utah. I. Title

QE169.F55 2000
557.92'5—dc21   99-087098

To my wife, Hilary ≈
   whose enduring patience and constant encouragement
   have helped to make this undertaking such a joy,

and to our son, Everett ≈
   may these lands remain a source of inspiration
   throughout your life.

And to the geology students of Western State College ≈
   who supply me with a constant barrage of humor
   and who tolerate me.

Contents

Introduction

First, let me inform you what this book is not. It is not a hiking guide to southern Utah. It will not direct you to splendidly isolated canyons, mostly because they would no longer be splendid or isolated. This book is designed to enhance your visit to any part of southern Utah by describing and interpreting the geologic events that have shaped this fantastic country. It is my feeling that an understanding of the geologic events that formed the rocks and landscapes of a region make any visit a much richer experience.

The seed of this book was sown in the summer of 1978 when I moved from the Midwest to a weird little town on the west slope of the Colorado Rocky Mountains. Soon after, I discovered an article by Edward Abbey in the late, great "magazine" *Mountain Gazette* that pulled me to southern Utah. I was stunned on that rainy fall morning by the bizarre fins and towers, the heights and the depths. I had found a place—the southern Colorado Plateau—that would take more than a lifetime to explore fully. Clearly, I had some catching up to do.

My friends at this time were a ragtag bunch of climbers and skiers, several of whom were studying geology at Western State College, where a profound lack of interest had doomed my earlier attempt at higher education. Their stories of landscape evolution and the origins of these rocks made my head swim and filled me with a zeal that was previously reserved only for climbing. In the rocks of western North America, those building blocks of mountains and canyons, I had discovered what is best termed an obsession. Obviously, geology was the great motivating influence I needed to succeed in college. My immediate goal was to understand the evolution of this fantastic western landscape. Thus began a journey without end, the kind I like.

After ten years of college culminating in a Ph.D. in geology from the University of Kansas, I returned west to Flagstaff—to live on the Colorado Plateau again and eke out a living as a temporary instructor at Northern Arizona University. In 1996, while I was at NAU, the seed of this book took root. In September I heard rumors of a major announcement and ceremony at the Grand Canyon in which President Bill Clinton would declare a new national monument. People talked of the Escalante River drainage and the Kaiparowits Plateau. This was thrilling news, as I had long considered this region to be the most beautiful and isolated on the Colorado Plateau, but also the most threatened by development.

I attended the announcement, held on the south rim of the Grand Canyon with the great chasm as the backdrop. The reality was even more fantastic than the rumors: a new national monument stretching westward from Capitol Reef all the way to the borders of Bryce Canyon! The day was incredible, thanks to inspiring speeches by Charles Wilkinson and a passionate reading by Terry Tempest Williams that drew tears from many in the crowd. By the time Vice-President Al Gore and President Clinton reached the stage, the crowd was in a celebratory mood.

On the drive back to Flagstaff that day I resolved to do something for this part of the great plateau. The next week I began writing the book you hold in your hand, two years of the most enjoyable "work" I have ever undertaken.

During these two years of writing, incredible things have happened, including the birth of a son, Everett. There is nothing like having a child to convince any sane person of the need for the preservation of wild places for future generations. Additionally, I became a permanent faculty member in the Geology Department at Western State College in Gunnison, Colorado, where much earlier I had become so enamored with the subject. So here I sit at the end of the semester, planning a two-week foray into the southern Utah wilderness. While some things are always changing, some, hopefully, will never change. I hope you enjoy reading and using this book as much as I have enjoyed writing it.

The Book ≈

This book is divided into two parts for ease of use. The first begins with a brief introduction to the general principles of geology, including the various rock types and their genesis, rock deformation (folds and faults), and geologic time. Following this whirlwind tour (half a semester in Geology 101!) we plunge into the geologic history of southern Utah. This is organized as a geologist deciphers the earth's record—chronologically, beginning with the oldest rocks exposed in the region and ending with modern geologic processes. The oldest rocks exposed in the area are of Permian age, between 286 and 245 million years old. Thus the record before 286 million years is ignored here. The geologic record of the Grand Canyon ends where this book begins. It is written this way for two reasons. First, it focuses on southern Utah; older rocks are definitely present in the subsurface, but can not be seen as one travels across the area. There is no sense in describing these rocks for the general public if nobody sees them. Second, plenty of very good books are now available on Grand Canyon geology that do discuss these older rocks. There is no reason to repeat that information.

The second part of this book consists of easy-to-use geologic road logs with mile by mile explanations of geological features. It is organized from east to west and mostly covers the state highways that traverse the region. The logs are broken into chapters, beginning in the town of Hanksville at the foot of the Henry Mountains. From there the route passes through Capitol Reef to the town of Torrey and southward through Boulder and Escalante into the heart of the new Grand Staircase–Escalante National Monument. As it continues west, the route cuts across the Kaiparowits Plateau and then climbs to Bryce Canyon. After Bryce, it runs across the High Plateaus and south through Zion National Park. The road log terminates near the town of Hurricane, which lies on the west edge of the Colorado Plateau. The logs are keyed to odometer mileage and prominent landmarks. In addition to the geological interpretations, a limited amount of historical information is included, especially with regard to the early Mormon settlement and the exploratory expeditions led by Major John Wesley Powell. His surveys figure prominently in the regional history. It was in southern Utah that many early principles of geology were developed, thanks to his associates G. K. Gilbert and C. E. Dutton.

In part II I try to be thorough in my explanations. This results in a certain amount of overlap between the two parts, but also allows them to be used independently of each other. It may be helpful, however, to flip back to part I when interested in more detail on a specific rock unit or feature, particularly to the regional paleogeographic (ancient geography) maps for that time.

Finally, I have tried to use the most up-to-date references available for this region. The science of geology is, and always will be, a work in progress, and many articles are published yearly on the geology of this area. For more detail on a particular aspect of the geology of southern Utah, the reader is referred to the articles and books listed at the end of each chapter. I owe a huge debt to the geologists listed in these references. Their research and the groundwork they established allowed my own research for this book to progress smoothly.

Sketch of south-central Utah looking north from the Utah-Arizona border, showing topographic features, including the various plateaus, mountain ranges, and canyons of the region. From Gregory and Moore 1931.

1 ≈

Geologic History
of Southern Utah

1

Basic Principles of Geology

The Present as the Key to the Past ≈

Geology differs from other sciences in that time is a huge, almost over-whelming factor in all aspects of the subject. We are not physically able to go back 150 million years to float down the turbulent rivers that deposited the sandstone and conglomerate of the Morrison Formation. Nor would most people want to, considering the assortment of dinosaur bones recov-ered from these Jurassic rocks. Instead, geologists gather all the informa-tion available from the rock record, including composition, textures, and structures in the rock. These scraps of information are then pieced together to produce the clearest, most accurate picture possible. From this partially reconstructed puzzle, we must *interpret* the physical setting for these long-ago times. Interpretations are backed up by observations of modern geo-logic processes and their products.

A few years ago I taught a course on sedimentary rocks at Northern Arizona University in Flagstaff. One weekend we drove north to the trad-ing post/community of Cameron, where the highway crosses the Little Colorado River. In the moist sand of the dry riverbed we used shovels to dig trenches, constructing cross sections of the riverbed that revealed an array of ripples and dunes created by sand the last time the river was in flood. We could see the incline of the sand layers (called crossbeds), with their slope direction corresponding faithfully to the downstream current di-rection, just as the textbook had said. We then turned to the sandstone cliffs of the Chinle Formation that had been looming over us as we toiled. In these 225-million-year-old deposits were the same geometric forms of ripples and dunes, crossbeds formed by the currents of ancient northwest-flowing rivers engraved into the rock record.

This fieldtrip not only taught us about modern and ancient river de-posits, but illustrated one of the most fundamental tenets of geology, the concept of *uniformitarianism*. This long-winded term refers to the princi-ple that the natural processes observed in action today, things like floods, volcanic eruptions, climate changes, and extinction, are the same that have occurred throughout the geologic past. Thus we can compare the products of modern processes to those in the rock record and infer past activity. Of course, some things have evolved through time and are not the same—such as the types of life on earth and the chemistry of the atmosphere and oceans. Still, erosion has whittled mountains down to plains and rivers have carried mud, sand, and gravel to the sea since the earliest days of earth, just as they do today. Regardless of what the creature of the moment happens to be—and many have come, dominated, and disappeared—it is the physical processes that continue to endure.

The Rock Cycle ≈

Minerals ≈

The most fundamental aspect of geology is rock type. Before looking at rocks, however, we must first learn what they are made of. Rocks are composed of minerals—one type or more commonly several types in variable amounts. Some of the more common rock-forming minerals include quartz, feldspar, and the mica minerals, biotite and muscovite. In reality there are more than 2,000 known mineral types around the world, but only a few occur in real abundance to form the common rocks.

A mineral is defined by its distinctive chemical composition and crystal structure. Quartz, for instance, the most common mineral on earth, always has the chemical ratio of one atom of the element silicon (Si) to two atoms of oxygen (O), giving it the chemical formula SiO_2, one of the simplest formulas. In comparison, potassium feldspar, the next most common mineral, has the formula $KAlSi_3O_8$, meaning that for every three silicon and eight oxygen atoms it contains one potassium (K) and one aluminum atom. In any particular mineral, the atoms that make it up are always arranged into the same geometric shape, giving it the crystal form that is unique to that mineral. The form may not always be reflected by recognizable crystal faces, but rest assured that it is there on a very small scale.

Most minerals have a similar root. They crystallize from a chemical soup of magma, which, when cooled, forms igneous rocks (fig. 1.1). *Magma* originates at great depths in the earth. Wherever the temperature approaches 1200° C preexisting rock is melted, forming magma. Such places are not randomly scattered across the globe, but are controlled by interactions among the various tectonic plates that make up the oceans and continents. Upon melting, the buoyant magma slowly passes up through the cooler crust. This deliberate ascent is analogous to the rise of a hot-air bal-

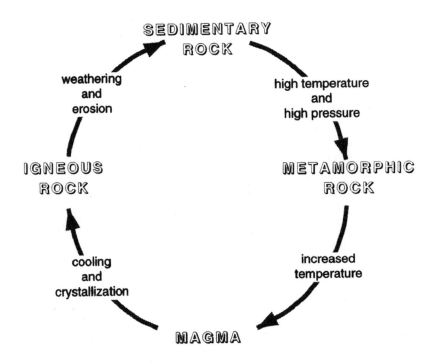

Fig. 1.1. The geologic rock cycle showing the relationship between the three main rock types and the processes involved in forming them. Note that it is not necessary to go all the way around the cycle to change one rock to another type. For instance, to obtain a sedimentary rock from a metamorphic rock, the metamorphic rock does not have to turn to magma, form an igneous rock, and then break down into sediment. Instead, a metamorphic rock at the earth's surface can be reduced to sedimentary particles through weathering and erosion to form a sedimentary rock directly. Similarly, an igneous rock can be subjected to intense heat and pressure, changing it into a metamorphic rock.

loon. When the air in the balloon is heated, it expands. This decreases its density, which makes it more buoyant than the surrounding cooler air. The expanding hot air rises, pulling the inflated balloon skyward. As magma rises, it encounters increasingly cooler rocks. If the magma temperature remains high, it may melt parts of the surrounding rocks as it rises. If the magma cools as it rises, it will begin to crystallize.

Igneous Rocks ≈

Igneous rocks occur in two types, based on their textures and where they cooled. If magma cools slowly below the surface, where it is insulated by a thick blanket of overlying crust, a *plutonic* igneous rock results. The term "plutonic" derives appropriately from Pluto, the Roman god of the under-world. Plutonic rocks are characterized by a coarse-grained texture, mean-ing that all the mineral grains that make up the rock are readily seen with the naked eye. Granite is the most abundant of the plutonic rocks when chemical composition is considered. Most granites contain an interlocking mosaic of minerals, chiefly quartz, feldspars of various chemical composi-tion, and biotite. The coarse-grained texture is a product of the slow cool-ing process. The slow rate encourages the unattached elements in the liquid magma to seek out and bond with other compatible elements at an unhur-ried pace. Individual atoms are given more time to find their place and bond in an orderly manner, adding to the crystal structure of a mineral atom by atom. The larger the crystals are in a plutonic rock, the slower was its rate of cooling and crystallization. This principle can be extended to infer the relative depth at which crystallization occurred. In other words, the larger the grains, the deeper the rock crystallized in the crust. Now we begin to see how textures can be used to interpret the history of a rock.

The alternative path of magma is for it to push its way to the earth's sur-face to form a *volcanic* igneous rock. Upon reaching the surface, the magma is suddenly confronted with much cooler temperatures—going from at least 800°C to 0°C in a geologic instant. There the molten rock cools so rapidly that most of the loose atoms have no time to organize into co-herent mineral forms and are "frozen" in their random positions to form a glassy texture. Because there is no organized crystal structure, volcanic glass is not composed of minerals. Crystals that do occur in volcanic rocks are usually small and sparsely distributed; most of them grew in the earlier his-tory of the magma, while it was slowly rising through the crust. The re-maining liquid is quenched at the surface, producing the very fine-grained and/or glassy texture. In general, these igneous textures can be interpreted to indicate rapid cooling and a volcanic origin. Still, we must use care in using grain size as the sole line of evidence for the origin of igneous rocks. Texture is best used in conjunction with other features, such as local and re-gional relations with surrounding rocks.

When magma breaks the surface to form a volcanic rock, it may exit in any number of ways; the eruption style depends mostly on gas content and chemical composition. Gas-rich eruptions come from discrete cone-shaped vents (volcanoes) and tend to be explosive and destructive. In contrast, gas-poor magma produces fluid lava that may pour from large fractures or fis-

Igneous Rock Classification

	silicic	intermediate	mafic
plutonic (fine-grained)	granite	diorite	gabbro
volcanic (coarse-grained)	rhyolite	andesite	basalt

Fig. 1.2. Names used to describe igneous rocks of various origins and chemical compositions.

sures to spread in thin flows across the landscape. Chemical composition also enters into the equation, as the two end members have very different characteristics (fig. 1.2). Silica-rich or *silicic* magma tends to be viscous, which inhibits the release of gas. As it approaches the surface, the gas pressure builds to intolerable levels and it escapes violently, hurling bits of molten rock high into the air to form volcanic ash. The Colorado Plateau has not seen much in the way of silicic eruptions through its history, although huge volcanic centers of this type occupied its margins from 30 to 20 million years ago. Most notable are the Marysvale volcanic field along the western margin and the voluminous San Juan volcanic field that created the San Juan Mountains in southwest Colorado.

The other end of the chemical spectrum is *mafic* magma, which by definition contains high amounts of iron and magnesium. Mafic magma forms the common volcanic rock *basalt* (fig. 1.2). Lava of this composition is fluid, and gas escapes easily, making explosive eruptions a rare occurrence. Instead the lava flows freely to spread as thin layers over large areas. This results in extensive flat areas covered by resistant black basalt. The wide plateau of Boulder Mountain, whose bulk looms above the towns of Torrey and Boulder, is a typical basalt plateau. This vast flat-topped feature formed as large volumes of fluid basalt flowed over the landscape: the lava filled in all preexisting topography to form a volcanic plain. Because basalt is more resistant to erosion than the surrounding sedimentary rocks, millions of years of water and wind have carved the sedimentary rocks away, leaving the isolated basalt-capped plateau behind.

Sedimentary Rocks ≈

All rocks contribute to the raw materials required for the creation of sedimentary rocks, our next stop on the rock cycle (fig. 1.1). Most rocks originate in the subsurface, under high pressure and temperature conditions. When exposed at the earth's surface, they encounter a very different environment that makes them unstable, and they begin to break down. This slow disintegration produces the clay, sand, and gravel that eventually are buried and cemented to form the rocks shale, sandstone, and conglomerate, respectively. These are the *clastic* sedimentary rocks that dominate the rock record of the Colorado Plateau (fig. 1.3).

Water is a vital ingredient in the breakdown of preexisting rocks. The energy of moving water is a constant and dynamic factor in breaking rock into particles that can easily be transported in rivers. Even as sediment tumbles downstream, abrasion continues to work the particles, smoothing the sharp edges and further reducing the size.

Water also attacks the rock on a chemical level. It is especially effective at threading its way into the crystalline structure of the minerals and plucking out certain elements. This not only changes the composition, but also renders the mineral unstable and susceptible to further physical and chemical attack. Dissolved elements such as calcium and sodium are then easily swept to the sea or lakes, where they become concentrated. Eventually these elements recombine to form the *chemical* sedimentary rocks such as limestone ($CaCO_3$), gypsum ($CaSO_4$), and halite ($NaCl$), which is ordinary table salt. Most limestone is formed by marine organisms (clams, corals, etc.) that absorb the calcium from the water to produce biogenic $CaCO_3$ in the form of shells and other hard parts. Gypsum and halite form as water in which the elements are concentrated evaporates, leaving the elements to bond together to form *evaporite* rocks.

When sediment is deposited, it accumulates in horizontal layers called *beds* or *strata*, with each bed representing a discrete instance of deposition. The vertical change from bed to bed may reflect nothing more than a simple shift in the water or wind velocity, or it may express a profound change in climate and depositional environment. For instance, the sharp line that separates mudstone in the upper Chinle Formation from the overlying Wingate Sandstone throughout the Colorado Plateau marks a significant change in both climate and environment. Chinle mudstones represent the floodplain of sluggish rivers that meandered across a humid western North America. By contrast, the massive orange cliffs of the Wingate mark an extremely harsh and arid environment dominated by blowing sand across a regional blanket of constantly shifting dunes. Within the Wingate cliffs are horizontal lines that denote minor variations in wind velocity or direction,

Clastic Sedimentary Rock Classification

sediment size	mud and silt	sand	gravel
sedimentary rock	shale or mudstone	sandstone	conglomerate

Fig. 1.3. Names of common clastic sedimentary rocks and the sediment types that compose them. Note that increasing grain size can be used to infer an increase in the energy of the transporting current, whether the current is water or wind.

while sand dunes continued to dominate the landscape. The combination of beds and the lines that bound them produces the horizontal patterns referred to as *bedding* or *stratification*, a characteristic of sedimentary rocks. All sedimentary rocks exhibit stratification on a variety of scales. It is this distinguishing feature that allows them to be identified as sedimentary in origin, even from some distance.

Metamorphic Rocks ≈

When igneous and sedimentary rocks are exposed to high pressure and temperature, they become metamorphic rocks. If the word "metamorphic" is broken down into its components, the meaning becomes apparent. "Meta" means change and "morph" means form, referring to rocks that have been altered by heat and pressure while still in a solid form. These changes may be reflected in a variety of ways, depending on the original rock type and the amount of pressure and heat. For instance, when the sedimentary rock limestone is subjected to high temperatures, the originally fine particles grow into larger crystals to form the metamorphic rock marble. Chemically both rocks consist of $CaCO_3$, but original features in the limestone such as fossils or stratification are obliterated as the small grains recombine into large crystals (recrystallization).

When the sedimentary rock shale experiences high pressure and temperature, the tiny clay particles grow into platy crystals of mica. Metamorphic rocks dominated by the sheetlike mica minerals are called *schist*. Directed pressure on the minerals as they develop forces a preferred orientation to the crystals, so they grow parallel to each other in overlapping sheets. This parallel orientation of platy or needlelike minerals produces the metamorphic texture called *foliation*. The linear fabric that foliation imparts to metamorphic rock resembles sedimentary bedding from a distance, but upon closer inspection the coarse mica grains become apparent.

Metamorphic rocks are not exposed in southern Utah and, in fact, are rare over most of the Colorado Plateau. Their scarcity is due to the stable nature of the region, so that high pressure and temperature conditions have not occurred for more than 1 billion years. Any metamorphic rocks present on the plateau are thus very old and deeply buried by a thick blanket of sedimentary rocks. The closest and most extensive exposures of such ancient rocks are found in the depths of the Grand Canyon, where incision by the Colorado River has laid bare the 1.7-billion-year-old Vishnu Schist. The chemical composition and remnant textures in these rocks indicate that, prior to metamorphism, the Vishnu was a thick sequence of sandstone, shale, and minor limestone with interbedded basalt and andesite lava flows. Similar metamorphic complexes likely form the basement of the rest of the Colorado Plateau.

Rock Deformation ≈

Rock, like any other solid matter, will deform when subjected to enough directed force or *stress* over some time interval. The exact response depends on many factors. A rock body will bend or *fold* under ductile conditions, for instance, when small stresses act over a long period (millions of years). Temperature will also affect the strength of a rock. If the temperature is high enough, the rock will fold under stress. Conversely, relatively cold rock at near surface conditions will behave in a brittle fashion, by fracturing or faulting. Faults are defined by a *fault plane*, the surface along which rocks on either side have moved relative to each other, either vertically or laterally. Faults have the effect of offsetting previously continuous bodies of rock (fig. 1.4).

The geometry of any rock deformation is controlled by the type of stress and the orientation of the forces that are involved. *Extensional stress* pulls the rock apart, lengthening but at the same time thinning the original body of rock. Extensional deformation stretches the rock and under brittle conditions will produce a *normal fault*. Normal faults occur when rock on one side of the fault moves down the slope of the fault plane as the rock is pulled apart (fig. 1.4A).

If stress is compressional—that is, the applied forces are pushing on the rock—it will thicken the body vertically and shorten it laterally. This is how mountains are uplifted. Under brittle conditions a *reverse fault* develops, where rock on one side of the fault is shoved up the slope of the fault plane (fig. 1.4B). A special type of reverse fault is the thrust fault, which forms when the angle of the fault plane is less than 30°, causing one side of the fault to be thrust up and over the adjacent block (fig. 1.4C).

A

B

C

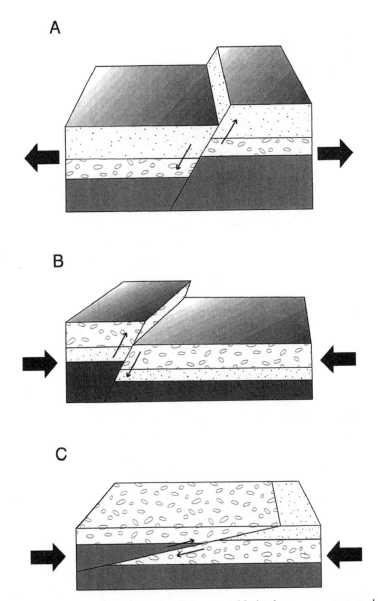

Fig. 1.4. Block diagrams showing the three types of faults that are common on the
Colorado Plateau or that may have played a role in its evolution. *A* is a normal fault
in which the down-dropped side moves down the slope of the fault plane. Normal
faults develop when the crust is pulled apart by extensional stress. *B* is a reverse fault
in which the block is pushed up and over the adjacent block. Movement is up the
slope of the fault plane and occurs under compressional stress. *C* is a thrust fault, a
special type of low-angle reverse fault where the slope of the fault plane is typically
less than 30°. Thrust faults typically form in layered sedimentary rocks during
extreme horizontal compressional stress.

Compressional stress and resulting thrust faults and folds have been the dominant mode of mountain-building in western North America over much of its geologic history. However, the stress regime has shifted in the past 30 to 20 million years to a dominantly extensional regime characterized by normal faults. The west margin of the Colorado Plateau has been especially affected by normal faulting where it is bounded by the Basin and Range province, a vast region deformed by large-scale extension. The stable interior of the Plateau has remained largely unaffected by this most recent deformation.

Folds are an important and often spectacular aspect of rock deformation. They form at all scales, from enormous mountain range–size folds to microscopic features. Compression is the most common mode of fold generation, but folds may form under any kind of stress. One simple requirement in the recognition of folds is layered rock. Without sedimentary bedding or metamorphic foliation to serve as a reference point, folded rock would be hard to identify.

Folds may take on a variety of shapes and forms; frequently several different types occur together. The two most common types are anticlines and synclines. *Anticlines* are uparched folds in which the layering slopes in opposite directions, away from a common central ridgeline called the fold axis (fig. 1.5A). The two opposite-dipping parts on either side of the axis are the *limbs* of the fold. *Synclines* are downwarped or troughlike folds with their axis located at the base of the trough (fig. 1.5A). The layering on each limb dips downward, toward the other limb and their mutual axis. Anticlines and synclines may occur in paired successions in which the limb of a syncline is shared with the adjacent anticline. Related successions of two or more folds are common in mountain belts, where sedimentary rocks are deformed by the intense compression generated during mountain-building. The axes of such related fold successions are aligned parallel to each other, but are perpendicular to the directed forces.

The most common fold type on the Colorado Plateau is the *monocline* or, as the name implies, single-limbed fold. Monoclines are large-scale step-like folds in which horizontal layering at a high elevation is linked to the same horizontal layers at a lower level by a single inclined limb (fig. 1.5B). These structures are the most striking features of the Colorado Plateau, with their great sandstone fins cocked skyward. The sinuous monoclinal ridges that snake across the landscape are cored at deeper levels in the crust by faults; overlying sedimentary layers are simply draped passively over these deep faults, providing a modified surface expression of the deformation. Where these faults have been exhumed by erosion, or viewed in the subsurface using modern subsurface imaging technology, the structures

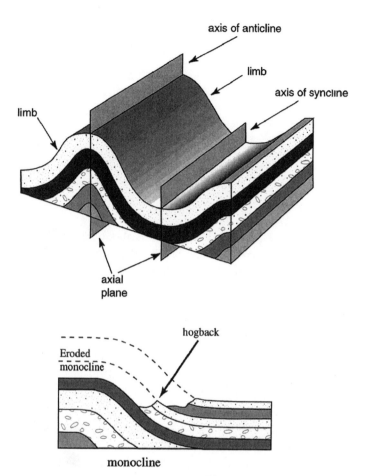

Fig. 1.5. Fold types commonly encountered on the Colorado Plateau. *A*: the block diagram shows an anticline-syncline pair. The anticline is the uparched fold on the left, while the downwarped syncline is on the right. Note that they share a limb. Each fold has an axis that bisects it into two limbs. Such a fold would be formed by compressional stress, oriented perpendicular to the axial planes. *B* is a cross section through a monocline or a single-limbed fold. Monoclines are by far the most common type of large-scale structure encountered on the Colorado Plateau.

have proven to be reverse faults, indicating an origin from compressional stress.

The large north-trending monoclinal flexures of the southern Colorado Plateau are obvious and spectacular features of the landscape that have long acted as barriers to east-west travel in the region. Even today roads traversing these structures are sparse and isolated. Major monoclines of the southern Colorado Plateau include the San Rafael Swell, the Waterpocket Fold that forms Capitol Reef, and the Cockscomb, a north-trending ridge that stretches northward from the Grand Canyon to form the west boundary of the remote Kaiparowits Plateau in the Grand Staircase–Escalante Monument.

Geologic Time ≈

Geologic time is a vast, almost incomprehensible concept when viewed from a human perspective, which is, after all, the only one we have. To consider 25-million-year-old rocks young takes quite a leap of the imagination. But when compared to the antiquity of the earth at 4.6 billion years (a billion is a thousand million), the rocks *are* young! If the same rocks are considered from the viewpoint of human origins about 1.8 million years ago, however, they are ancient. Geologists who deal with these quantities of time on a daily basis become practiced at filtering out the human perspective. Instead we learn to look at geologic time from a relative perspective, on a time scale relative to other rocks or geologic events.

Relative Ages and the Geologic Time Scale ≈
Even in geology's infancy as a science, time was recognized as an important element. The first attempts to organize rocks and the events that formed them into a coherent chronological order occurred in western Europe in the middle 1700s. As the science progressed, a more detailed chronology evolved. The geologic time scale used today largely has come from pioneering work done in the 1800s, although the actual number of years involved in geologic time was not realized until the 1950s. The geologic time scale was based on the *relative ages* of rock units recognized in western Europe. Relative age determination is based on the vertical relations between various rock units. Put simply, a sedimentary or volcanic rock unit is younger than the rocks below it, but older than those on top of it. As rocks were examined in ever greater detail, certain fossils were recognized as characteristic of specific units and could be taken to represent a discrete period. These observations allowed sedimentary sequences containing these fossils but separated by large distances to be correlated with respect to time; widely separated strata with the same fossils could be assumed to be the same age. These basic principles took years to be developed, tested for validity, and finally accepted by the geologic community; but they have withstood more than 200 years of testing and remain the basis for determining rock ages today.

In the early 1800s two British geologists, Adam Sedgewick and Roderick Murchison, set out to name the rock successions throughout Europe and determine their relative ages. The names established by them and others eventually came to represent formal units of time, which were used to construct the geologic time scale that today is recognized by geologists worldwide (fig. 1.6). Sedgewick and Murchison began this daunting task in Wales, where they distinguished two divisions of rocks. Lower rocks were designated the Cambrian System, after Cambria, the ancient Roman name for Wales. Overlying rocks became the Silurian System, named for the Silures, an ancient Celtic tribe that had inhabited the area. Earlier geologists had defined the Jurassic System for younger rocks in the Jura Mountains, which straddled the border between France and Switzerland. By the late 1800s many rock systems had been defined and their age relations resolved, giving birth to the relative time scale. The Cambrian System of rocks in Wales and its diagnostic fossil assemblage came to represent the Cambrian Period, a division of geologic time. It has only been in the last fifty years, with the advent of radiometric dating of rocks, that we have been able to determine an *absolute age* (in real numbers) for the Cambrian Period of 570 to 505 million years before present (m.y.b.p.). Prior to this important breakthrough, the age of rocks had to be established through diagnostic fossils contained in them. Such fossils continue to be important for age determinations in the absence of material suitable for radiometric dating. The principles behind radiometric dating are discussed in a later part of this chapter.

While the geologic periods were being defined, larger time divisions that included several periods also were recognized. These divisions, called eras, were based on broad groupings of the dominant fossil organisms for that time. The oldest is known as the Precambrian and encompasses 80 percent of geologic time, stretching from the beginnings of earth about 4.6 billion years ago to 570 m.y.b.p. This immense interval originally was defined by its lack of fossils. Recently, however, detailed study of late Precambrian sedimentary rocks around the world has turned up a unique fossil assemblage of strange, soft-bodied marine organisms. As a result, there is some discussion of revising the boundary between the Precambrian and the following Paleozoic Era, although no consensus has yet been reached.

The Paleozoic Era includes the Cambrian through the Permian Periods, a time span from 570 to 245 m.y.b.p. (fig. 1.6). Paleozoic means "ancient life"; this era began with the first appearance of shell-bearing organisms and their subsequent dominance of the marine environment. Life on land did not evolve until the middle of the Paleozoic. The following Mesozoic Era, to which most of the rocks of the Colorado Plateau belong, includes the Triassic, Jurassic, and Cretaceous Periods (fig. 1.6). The Mesozoic, which

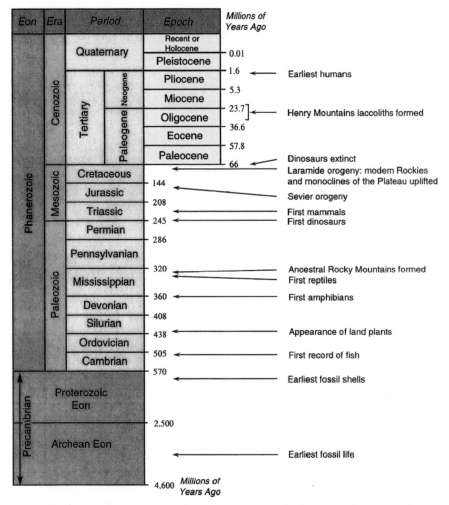

Fig. 1.6. Geologic time scale showing relative and absolute ages of eras, periods, and epochs and the dates of important events in the history of the earth and the history of the Colorado Plateau in particular.

translates as "middle life," was dominated by reptiles on land and in the seas, including the dinosaurs, whose extinction marks the end of the era.

The Cenozoic Era ranges from 66 m.y.b.p. to the present (fig. 1.6). The demise of the dinosaurs opened huge environmental niches for the mammals, who were patiently standing by awaiting the opportunity. As a result, the Cenozoic, which means "recent life," is typified by a diversity of mammals.

Radiometric Dating and Absolute Ages ≈

The geologic time scale, although derived from relative ages and fossils, shows real numbers or absolute ages at the period and epoch boundaries (fig. 1.6). These numbers have been determined through the painstaking radiometric dating of rocks at or near these boundaries. While the ages are accepted as accurate, this is an ongoing process in which the numbers are constantly being refined, either through more precision in the dating process or through the discovery and dating of new rocks at the various boundaries in the rock record.

The concept of radiometric age determination began with the discovery of radioactivity by Marie and Pierre Curie in 1896. It was not until 1907 that a method for dating rocks using natural radioactivity was proposed. The first time scale to include absolute ages was published in 1934.

Radiometric dating of rocks and minerals involves the radioactive decay of an unstable original atom, called the *parent*, to an atom of a different stable element, called the *daughter*. This process emits radioactivity. Several parent/daughter element sets are useful in radiometric dating. The more commonly used systems include uranium/lead (U/Pb), potassium/argon (K/Ar), and its more refined sister system, argon/argon ($^{40}Ar/^{39}Ar$), in which different states of the element are used. The system used to date any particular rock is dictated by the elements and minerals it contains.

When a mineral crystallizes from a melt of magma, atoms of radioactive parent elements become locked into the crystal structure and the radiometric clock begins ticking. The parent atoms decay to stable daughters in a regular and predictable way, in a measurable amount of time called the *half-life*. The half-life of any radioactive system is the time that it takes for half the parents to decay to daughters. This has been calculated through numerous experiments for all the systems used to date rocks. For instance, the half-life of the K/Ar and $^{40}Ar/^{39}Ar$ systems is 1,250 million years. In other words, over that period one-half of the parents decay to daughters. If the time since crystallization is two half-lifes, one-half the parents decay over the first half-life, leaving half the parents; over the second half-life, half the remaining parents decay to daughters, leaving only one-quarter of the parents that originally were present at crystallization. At this point three-quarters of the parent atoms have decayed to daughters. Using the measured parent/daughter ratio of that particular system in a particular rock and the known half-life, the time when the rock crystallized can be calculated. While in reality the entire process is not so simple as that, this is essentially how absolute ages are obtained for igneous rocks. The measurement of the ratios of specific elements in a mineral or rock requires a sophisticated and expensive piece of equipment called a mass spectrometer. The entire process is expensive and time-consuming, but ultimately worthwhile.

Although radiometric dating is limited to igneous rocks, the dates obtained from lava flows or volcanic ash that are interbedded with sedimentary rocks are valuable in refining relative ages in terms of absolute numbers. A sandstone unit might contain fossils that suggest an Oligocene or Miocene age, putting it in the rather wide age range of about 36 to 5 million years (fig. 1.6). However, the unit rests on a lava flow dated at 24 million years and is overlain by a volcanic ash layer that yields an age of 18.5 million years. Radiometric ages obviously provide much tighter constraints, placing the age of the sandstone between 24 and 18.5 million years, in the early part of the Miocene epoch. Radiometric dating has also proven effective in establishing the age range of fossils or groups of fossils in absolute terms. If the age range of a fossil assemblage can be determined using interbedded volcanic material, the presence of the same fossils in widely separated areas can be used to demonstrate similar age ranges in real numbers. In using these methods, the relative dating of rocks has grown increasingly precise, advancing to a level of refinement that was impossible before the advent of radiometric dating. Moreover, radiometric dating has profoundly enhanced our understanding of the dimension of time relative to earth history and the rock record.

2

Permian Period

Our journey into the geologic past of the magnificent parks and monuments of southern Utah is essentially a slow ascent of the Grand Staircase. This history begins with the oldest rocks exposed in the region, which are of Permian age. The Permian Period began 286 million years before present (m.y.b.p.), although this is not to imply that there is no earlier history. Quite the contrary! The geologic history of the Grand Canyon, which is incised into even deeper levels of the earth's crust a short distance to the south, reaches as far back as 1,750 m.y.b.p. While these older rocks have been documented in the subsurface of southern Utah, they take no part in the colorful and varied landscape that is our focus. The history recorded in the awe-inspiring strata of the Grand Canyon ends where that of southern Utah begins, in the Permian. Thus our story takes up where the Grand Canyon leaves off.

Tectonic Setting ≈

The Permian saw the final assembly of the supercontinent known as Pangaea ("all lands"), which included the continents of North America, South America, Africa, Antarctica, Europe-Asia, and Australia (fig. 2.1). As the continents joined together, collisions along their boundaries produced intense compressional stress, resulting in the uplift of linear mountain belts. Where the European landmass collided with eastern North America, the earth's crust was pushed and squeezed, giving rise to the Appalachian Mountains. To the southeast, where Africa and South America pushed into the North American continent, the Ouachita-Marathon orogenic belt formed in present-day Arkansas, Oklahoma, and Texas. These mountains formed a continuous belt that extended along the entire east and southeast margin of North America, although at this time they were a part of the vast interior of Pangaea.

An *orogeny* is a discrete mountain-building event that generally is confined to a specific geographic area, occurs within a definable time interval,

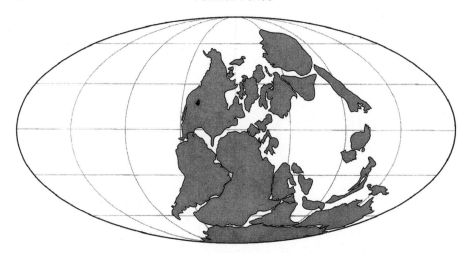

Fig. 2.1. The supercontinent of Pangaea during the Late Permian and Early Triassic, showing the location of North America and the state of Utah (in black).

and is characterized by a distinct style of deformation (folding and faulting). This informal but rather rigorous definition allows geologists to keep mountain-building episodes clearly separated in both time and space. However, as we will see, we are dealing with natural systems—and they do not always follow the rules! Orogenies typically are named for the geographic areas that best illustrate the deformation. For example, deformation associated with the Ouachita-Marathon orogeny is most clearly displayed in the Ouachita Mountains of northeast Oklahoma and northwest Arkansas. Here pre-Permian rocks are intricately folded and faulted by the unification of Pangaea. *Orogenic belts* are linear mountain ranges that are the expression of deformation. These mountain belts often are given the geographic name of the orogeny (e.g., Ouachita orogenic belt). Normally, orogenic belts are elongate perpendicular to the dominant stress direction (fig. 2.2). An analogy would be crumpling a flat-lying piece of paper by applying horizontal pressure to it. Wrinkles or ridges in the paper form at right angles to the direction of pressure. This is how the earth's crust behaves. By studying the orientations of folds in orogenic belts, geologists are able to determine the orientation of ancient forces and the direction from which they were generated.

Our concern is the western interior of North America, which centers on the Four Corners states of Utah, Arizona, New Mexico, and Colorado as well as the states that bound them. The most prominent Late Paleozoic tectonic feature in this region was the Ancestral Rocky Mountains, located in western Colorado and New Mexico (fig. 2.3). Uplift of these mountains coincided with the Appalachian and Ouachita-Marathon orogenic belts,

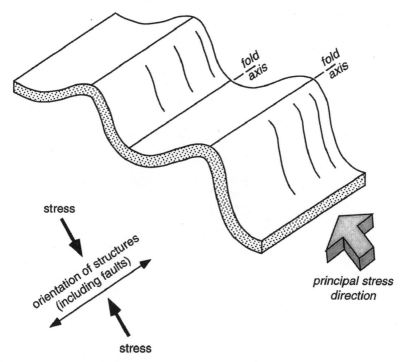

Fig. 2.2. Diagrams showing the orientation of directed stress relative to the orientation of resulting structures such as folds and faults. Note that the axes of the folds and fault trends form perpendicular to the directed stress.

beginning during the earliest part of the Pennsylvanian Period and continuing into the Permian. The mountains remained a positive feature through part of the subsequent Mesozoic Era, influencing regional drainage patterns and sedimentation for tens of millions of years. Although these mountains occupied the site of the present-day southern Rockies (hence their name), erosion had planed them flat long before the more recent uplift to their present stature.

The Ancestral Rockies in western Colorado were composed of two parallel northwest-trending ranges separated by a deep, narrow basin (fig. 2.3). The eastern belt is called Frontrangia (or ancestral Front Range) for its location that coincides with the present-day Front Range, which presently forms the western backdrop for the Denver-Boulder area. The western mountain belt was situated along the Colorado-Utah border and is called the Uncompahgre highlands. It was these western highlands that profoundly affected the style and amount of deposition in southern Utah through its Late Paleozoic and Mesozoic history.

The driving mechanism for uplift of the Ancestral Rockies has recently

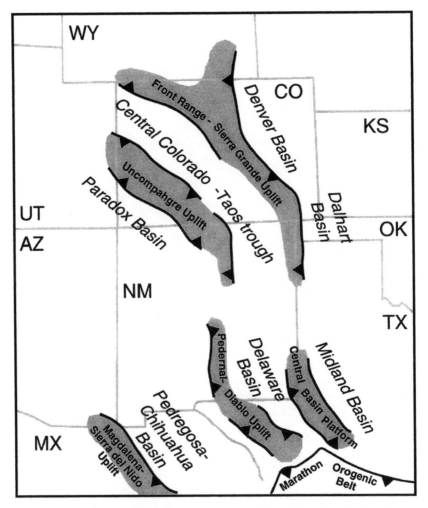

Fig. 2.3. Permian geography of part of southwestern North America, showing location of the uplifts (shaded), the faults that bound them, and the associated basins. Barbs are on the uplifted side of the bounding faults.

become the source of controversy. It has long been thought that uplift was related to the Ouachita-Marathon orogenic belt and continental collision along the southeast margin of North America (fig. 2.4). In this scenario, compressional stresses along the southeast margin developed as South America crowded into the continent. These stresses were transmitted far inland to push up the Ancestral Rockies in New Mexico and Colorado (Kluth 1981). If the Ancestral Rockies are indeed the result of the Ouachita-Marathon event, they did not behave properly! In fact, faults along which the Ancestral Rockies were uplifted trend northwest, *parallel* to the north-

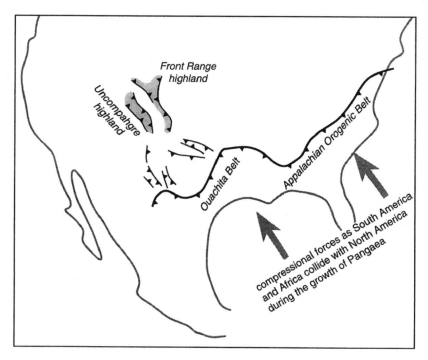

Fig. 2.4. Map of North America showing the geometric relationship between the northeast-trending Ouachita-Appalachian orogenic belt formed during the construction of Pangaea and the northwest-trending tectonic elements of the Ancestral Rocky Mountains, specifically the Front Range and Uncompahgre highlands in western Colorado (shaded). After Ye and others 1996.

west-directed compression determined from the Ouachita belt (fig. 2.4). They should instead trend northeast-southwest. This puzzling geometry has been explained by the presence of preexisting northwest-trending weaknesses in basement rocks of the western interior, possibly formed during a much older orogenic event. "Basement rocks" or simply "basement" refers to a complex of igneous and metamorphic rocks that form at deep levels in the earth's crust and serve as the foundation on which the numerous sedimentary layers of rivers and seas are laid down. Northwest-trending faults in the basement of western Colorado and New Mexico are thought to have been reactivated as compressional stresses pulsed through the crust from the southeast.

As in all physical sciences, ongoing geological research drives constant advancements in the understanding of our planet. New studies may cause older interpretations to be refined or may spark alternate ideas. One recent study suggests that uplift of the Ancestral Rockies is in fact unrelated to the Ouachita-Marathon orogeny. According to Hongzhuan Ye and others

(1996), an alternate hypothesis may be tied to compression related to a subduction zone located along the southwestern margin of North America. Before the implications of such a setting can be adequately discussed, however, a basic understanding of the mechanics of plate tectonics is necessary.

Plate Tectonics: A Primer ≈

Two kinds of plates make up the earth's crust, that outer rind on which we live. *Oceanic* and *continental plates* differ from each other in almost all aspects—in fact, the only feature they share is that both plates are mobile and, geologically speaking, are constantly changing.

Oceanic plates are found (surprise!) beneath most oceans. *Oceanic crust* is relatively thin, less than 10 kilometers thick. Basalt, a dense black volcanic rock, makes up most oceanic crust. High concentrations of iron and magnesium give basalt its high density; because of these chemical components, it is classified as a mafic volcanic rock. Fresh basaltic ocean crust is constantly being generated by volcanic activity in the deep ocean at what are known as *mid-ocean ridges* (fig. 2.5).

Mid-ocean ridges (MORs) are classified as *divergent plate boundaries,*

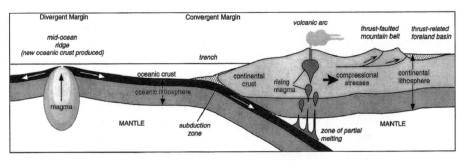

Fig. 2.5. General cross section through the crust and upper mantle of the earth showing relationships between the various types of tectonic plates and their boundaries as well as the geographic features that form as a result of these plate interactions. This particular cross section is analogous to the modern plate tectonic setting in the Pacific Ocean and South America. The analogous mid-ocean ridge is the East Pacific rise, with the left side representing the Pacific ocean plate and the right side the Nazca plate, which is being subducted beneath the South American continent, immediately west and offshore of the continent. The trench, which is the ocean floor expression of the subduction zone, is the present-day Chile-Peru trench. The volcanic arc and thrust-faulted mountains would be analogous to the modern Andes Mountains, which form the western borderlands of the South American continent. A similar situation existed along the west margin of North America during the Jurassic and Cretaceous Periods.

one of three types of interactions that are possible wherever two tectonic plates are in contact with each other (fig. 2.5). Mid-ocean ridges separate two discrete oceanic plates in places where the crust is literally being pulled apart. Fractures that open up along these boundaries allow molten rock, called magma, to well up from deep within the crust and spill out onto the ocean floor. Here contact with cold ocean water causes instant solidification into basalt. In this manner new oceanic crust is produced.

As the crust continues to pull apart or extend, the newly formed crust splits: the two segments move in opposite directions, as parts of two different plates. Thus, a continuous cycle of crustal production and plate movement occurs.

Although most MORs are located in the middle of ocean basins, the higher temperature of the young rock causes them to be elevated above the surrounding cooler and denser crust. Mid-ocean ridges are characterized by continuous, narrow undersea mountain chains that, when viewed on a global scale, resemble the stitching on a baseball, although certainly not in such a regular pattern. Presently both the Atlantic and Pacific Ocean basins are split by MORs.

As oceanic crust moves conveyor belt–style away from its MOR origin, it gradually cools, becomes denser, and begins to sink to a lower elevation. Since new oceanic crust is constantly being produced, and the earth is not expanding to accommodate all this added crust, it must somehow be disposed of or destroyed. The solution to this problem lies in the relationship with continental crust and the second type of plate boundary.

In contrast to oceanic crust, *continental crust* ranges from 30 to 70 kilometers thick. Continental crust is relatively deficient in the elements iron and magnesium and instead is rich in silicon, so that it is less dense than its oceanic counterpart. Continental crust consists dominantly of granite, an igneous rock that contains an abundance of the silica-rich minerals quartz, feldspar, and mica.

The second type of plate boundary is the *convergent boundary*. This type occurs wherever two plates push against each other and is characterized by compressional stress and mountain-building. While several combinations of plate interactions are possible, the most common type is that between oceanic and continental plates. Where this type occurs, dense oceanic crust is shoved beneath the relatively buoyant continental plate, a situation known as *subduction* (fig. 2.5). Put simply, the oceanic plate is subducted beneath the continental plate—and, as any geologist can tell you, subduction leads to orogeny!

Two significant geographic features are the direct result of subduction: volcanic mountains and compressional orogenic mountain belts. As the oceanic plate is subducted, it dives to increasingly deeper levels in the crust. Because the temperature in the earth increases with depth, the subducting plate eventually reaches a zone in which it begins to melt (fig. 2.5). As magma is produced from the melting plate, density contrasts between materials again become a controlling factor in their behavior. Since hot magma is less dense than surrounding cooler country rock, the magma rises slowly through the crust, much as a hot-air balloon ascends through the cooler atmosphere. If magma finds its way to the surface, a volcanic eruption takes place. If not, it cools slowly at shallow levels in the crust to form a coarsely crystalline plutonic or *intrusive* igneous rock, such as granite. Volcanic mountain belts that form inland of subduction-related continental margins are called *volcanic* or *magmatic arcs*. The nearest modern example of an active, subduction-related volcanic arc is the Cascade volcanoes in the Pacific Northwest states of Oregon and Washington. Here the Nazca plate (oceanic) is subducting eastward beneath the continental North American plate. In fact, all of the modern Pacific Rim volcanic activity, including South and Central America, Mexico, Alaska, and Japan, is related to subduction as oceanic crust is consumed and recycled into continental crust.

Another feature of subduction-influenced continental margins is orogenic mountain belts formed by compressional stresses that are generated during subduction. As an oceanic plate is consumed beneath a continent, huge amounts of compressional stress are transmitted horizontally through the overlying crust, thrusting toward the interior of the landmass. This results in folding and faulting of the crust, which is expressed at the surface in the form of mountain belts. Often the remnants of volcanic arcs and orogenic belts are the only evidence for an ancient subduction-dominated setting. It is this type of evidence that forms the basis for the recently proposed origin of the Ancestral Rocky Mountains.

Meanwhile, Back at the Ranch . . . ≈

There are several reasons why the Ouachita-Marathon orogeny has long been invoked as the catalyst for uplift of the Ancestral Rocky Mountains. First, the two events are approximately coeval. When looking for a relationship between two geologic features, one of the most obvious similarities is the timing of the events that formed them. Second, there was no other apparent source for compression of the magnitude required for such an uplift. This is due in part to the dearth of information regarding the Late Paleozoic history of Mexico. Considering the northwest trend of structures in the Ancestral Rockies, the direction from which the compression should have come was the southwest, possibly placing the source in Mexico.

Indeed, the alternate hypothesis of Ye and others (1996) points to a source for compression in the southwest, rather than the Ouachita-Marathon belt. This is based on recent studies in Mexico and fits well with the orientation of faults, folds, and basins associated with the Ancestral Rockies (fig. 2.4). J. W. McKee and others (1988) recognized a previously unknown belt of Pennsylvanian and Permian volcanic rocks in east-central Mexico that probably are the remnants of a volcanic arc, implying a nearby subduction zone. This inferred convergent boundary has been called upon as the driving mechanism for uplift in Colorado and New Mexico. Further evidence for this association includes similar Pennsylvanian-Permian north-west-trending uplifts and basins in northern Mexico. These structures, which likely are related to the Ancestral Rockies, are too far south to have been generated by the Ouachita-Marathon event. Although much work remains to be done on the complexities of the geology of Mexico, the age and orientation of the Ancestral Rockies certainly fit with a convergent plate boundary in Mexico during this time.

Geologic Setting of the Permian Colorado Plateau ≈

As we have seen in the rather long-winded but necessary discussion of the Late Paleozoic tectonic setting, the Permian history of the Colorado Plateau actually began in the previous period. During the Pennsylvanian Period, uplift of the Uncompahgre highlands not only provided a highland source for clastic sediment to be shed west, but also had a great influence in the subsidence of the adjacent Paradox basin. The Paradox basin was a deep, northwest-trending trough that bordered the Uncompahgre highland along its southwest margin (fig. 2.3). Marine deposits dominated the region except along the edge of the highland, where a fringe of coarse sand and gravel collected. High subsidence rates driven by the adjacent Uncompahgre uplift resulted in the accumulation of more than 2,000 m of marine sediments in the trough. The basin shallowed farther to the southwest, away from the highlands. Thick beds of halite (ordinary table salt) and other evaporite minerals that were deposited during this time mark periodic restrictions of the seaway. When fresh seawater was obstructed from entering the basin, dissolved salts became concentrated. Evaporite minerals were precipitated when evaporation rates exceeded the influx of fresh water.

The thick salt deposits of the Paradox basin have been an important factor in the later geologic history of east-central Utah. When salt is put under pressure (for instance, by the weight of subsequent sedimentary layers), it flows plastically in an attempt to escape the overlying force. It may move laterally until it encounters a weakness, allowing it to migrate vertically. Upon finding a weakness, such as a fault or a zone less encumbered by overlying layers, the less dense salt begins to rise, pushing through the surrounding rock, much as magma rises through more dense crust. As the salt forces its way through the overlying layers, they are folded upward. Upon breaching the surface, salt is rapidly dissolved, and valleys are easily carved. Typically, the sedimentary layers on either side of such a valley tilt away from the valley's axis. Uparched folds formed by the movement of salt to the surface are called *salt anticlines*. Modern salt anticline valleys in east-central Utah include Castle Valley, Spanish Valley (in which the booming town of Moab is located), Salt Valley in Arches National Park, and the Paradox Valley in western Colorado, to name but a few. All the salt anticline valleys associated with the Paradox basin trend northwest, suggesting that even the movement of salt was controlled by preexisting northwest-trending weaknesses.

By the beginning of the Permian the sea had receded westward, exposing the uppermost Pennsylvanian strata to the forces of erosion. A variety of continental and marine systems occupied the Colorado Plateau region and resulted in a bewildering array of deposits. Through most of the Permian continental conditions existed in eastern Utah, while the western part was covered by a shallow seaway. These contrasting environments were separated by a constantly shifting north-trending shoreline. Sea level fluctuated considerably, due to the combined effects of real global sea-level changes because of the waxing and waning of polar ice caps and local variations in subsidence and sediment supply. The Permian Period in the western interior, like the Pennsylvanian before it, was arid, and this is reflected in both marine and continental deposits.

In an east-to-west direction across the Colorado Plateau a distinct facies pattern can be discerned in Permian strata (fig. 2.6). "Facies" is a term for a particular "package" of rock that represents a specific depositional environment (e.g., river, lake, sand dunes) that is distinguished by, among other things, its rock type, sedimentary structures, grain size, fossils, and sometimes color. For example, "eolian facies" refers to windblown sand-dune deposits that are characterized by fine and medium-grained sand (sand is defined as particles up to 2 mm in diameter) with large-scale crossbedding (usually more than 0.5 m thick). Fossils are rare because of the extreme climatic conditions that prevail in this type of environment. An arid climate is

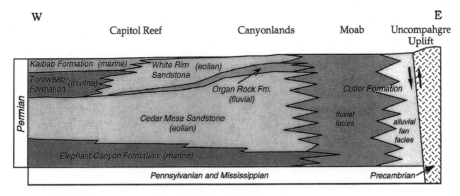

Fig. 2.6. Diagrammatic cross section through southern Utah showing depositional settings and relations of the various Permian units in this part of the Colorado Plateau. After Blakey 1996.

implied because wet or damp sand is not easily moved by wind. Thus, facies are not only useful as a general description for a body of rock, but also may provide information regarding paleoclimatic and paleogeographic conditions.

Throughout the Permian powerful streams cascading westward from the Uncompahgre highlands carried coarse debris into the subsiding Paradox basin. These coarse proximal sediments, containing particles up to boulder size, form the Cutler Formation. The Cutler Formation was deposited in apronlike alluvial fans at the base of the mountains and in the west-flowing rivers that threaded their way seaward. Grain size in these stream-laid sediments diminished westward, as the turbulent rivers that flanked the highlands gave way to tranquil streams of the low-lying coastal plain. The gradient and energy of the Permian rivers, and thus their ability to carry coarser sediment, decreased with distance from their mountainous source. Farther west the Cutler grades into a variety of facies. The adjacent environment largely depended on sea level and sediment supply.

The lowermost Cutler Formation grades westward into the Elephant Canyon Formation, a unit dominated by fossiliferous limestone, but also containing variable amounts of shale and sandstone, and locally, conglomerate (fig. 2.6). Abundant and varied fossils indicate a shallow sea teeming with marine organisms. Frequent flood events in Cutler rivers flushed coarse sand and gravel into the sea, to be reworked by waves and marine currents along the shoreline. The Elephant Canyon Formation is magnificently displayed along the canyon walls at the confluence of the Green and Colorado Rivers in Canyonlands National Park.

Succeeding the Elephant Canyon Formation are eolian deposits of the

Cedar Mesa Sandstone (fig. 2.6). Strong northwesterly winds ferried large quantities of sand into eastern and south-central Utah, shaping it into large dunes that pushed the sea farther westward. The Cedar Mesa Sandstone is widely exposed in the canyon country of southeast Utah, in Canyonlands National Park, Natural Bridges National Monument, the Dark Canyon area, and capping its namesake, Cedar Mesa. The sandstone forms the striking red and white striped pinnacles and mushroomlike towers that characterize the Needles district of Canyonlands. Correlative eolian deposits exposed in the upper walls of the Grand Canyon constitute the Esplanade Sandstone.

The vast coastal Cedar Mesa dune field was blanketed by fine-grained red beds of the Organ Rock Formation (fig. 2.6). These beds of shale, siltstone, and sandstone are the distal equivalents of the coarse Cutler Formation that continued to be shed off the highlands in Colorado. The Organ Rock Formation is the product of rivers that wound lazily across the vast, featureless floodplain to their western shoreline along the Utah-Nevada border. Lens-shaped sandstone bodies floored by conglomerate are the preserved channels of these ancient rivers. Conglomerate is intraformational in origin, as the rivers lacked the vigor to transport pebble-sized sediment this far from their Uncompahgre headwaters. "Intraformational" means the sediment was derived from erosion of material within the formation, in this case partially consolidated floodplain deposits. As the rivers wandered back and forth across the floodplain, clumps of cohesive muddy sediment fell into the channel; their pebble-sized remnants eventually came to rest at the bottom of the channel. Like previously discussed Permian units, the Organ Rock Formation is not exposed in the park lands of south-central Utah but is well displayed in the extremely dissected country of Canyonlands National Park, a short distance northeast, as well as Monument Valley along the Utah-Arizona border. The Organ Rock Formation is the equivalent of the Hermit Formation in the upper reaches of the Grand Canyon, which has a similar origin.

White Rim Sandstone and Kaibab Formation ≈

The White Rim Sandstone overlies the Organ Rock Formation and marks the return of widespread eolian dunes to the region (fig. 2.6). The White Rim Sandstone and its marine counterpart to the west, the Kaibab Formation, are the oldest exposed rocks in the south-central Utah park lands. In Capitol Reef they can be viewed in the towering walls carved by the Fremont River where it slices through Miners Mountain. A short distance to the south, these rocks are exposed in the Circle Cliffs area just east of the labyrinth of canyons formed by the Escalante River drainage. Uplift of the Circle Cliffs began about 60 million years ago, allowing the forces of erosion to peel back younger strata, laying bare the White Rim Sandstone and Kaibab Formation.

The walls of the Fremont River canyon tell the story of the interactions between the encroaching sea and the vast dune field, for it is here that the Permian shoreline lay. Although the White Rim Sandstone is dominated by large-scale crossbedding typical of eolian dune deposits, thin beds sandwiched within show evidence of marine flooding. The most obvious features are the dwelling and feeding traces of marine organisms, which erased any preexisting sedimentary structures. Additionally, thin lenses of limestone formed in low-lying areas between the large dunes as the sea invaded the west margin of the dune field.

As the sea advanced from the west, sand that piled up into dunes blew in from the north. This is evident from crossbedding in the ancient dunes that dips to the south and southeast, recording prevailing winds in those directions. The preservation of sedimentary structures like crossbeds is vital to geologists because they form in response to wind or water currents. Crossbeds in eolian deposits tell us the dominant wind direction, which is useful in the reconstruction of ancient climates. Crossbedding in streamlaid sediment indicates the direction the river flowed. Combining this with other information, such as the mineral composition of the sediment, may reveal the ancient source mountains. Geologists collect all these scraps of evidence to obtain the most accurate picture possible of ancient landscapes.

While the White Rim dune field spread over a wide belt of the late Permian coastline, powerful winds blew sand eastward over the low-lying fluvial plain. Although sand along this margin never accumulated in quantities sufficient to evolve into dunes, an extensive sandflat did develop (fig. 2.7). West-flowing rivers worked their way across these sandflats and the dune field on their seaward journey. The passage of these rivers is evident in their coarser channel-shaped deposits amid crossbedded dune deposits.

A gradual sea-level rise drove the sea eastward, depositing basal marine sands of the Kaibab Formation over the White Rim dunes (fig. 2.7). As the water invaded and reworked the dune tops, marine organisms colonized the sandy, shallow-water environment. A diverse fossil assemblage, including abundant brachiopods, gastropods, bryozoans, and crinoids, indicates a healthy fauna and normal marine conditions. With time, the production of calcium carbonate by these organisms escalated to the point that limy sediments overwhelmed the sand supply. Much of the resulting sediment was rich in the carbonate mineral *dolomite*, which forms when half the calcium

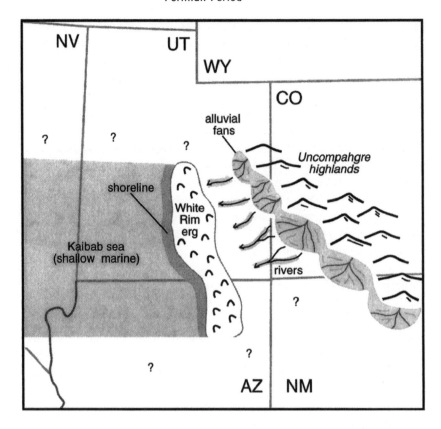

Fig. 2.7. Paleogeography during deposition of the Permian White Rim Sandstone and the Kaibab Formation on the Colorado Plateau. After Condon 1997.

ions in calcium carbonate (normal limestone) are replaced by magnesium ions. Calcium carbonate, which is both the mineral calcite and the rock limestone, has the chemical composition $CaCO_3$, whereas dolomite, also a mineral and rock, has the composition $MgCa(CO_3)_2$. This replacement in the Kaibab Formation probably occurred when magnesium-rich seawater reacted with the biochemically derived calcium carbonate.

Limestone in the Kaibab Formation contains abundant chert in the form of irregular nodules or thin layers. Like the mineral quartz, chert is composed of silicon and oxygen (SiO_2), but it has much smaller crystals. In most cases chert forms after limy sediments have been cemented or *lithified* into limestone. As we have seen, calcium carbonate is extremely reactive with fluids that come into contact with it. When naturally acidic waters (as most are) pass through fractures or pores in the limestone, it is dissolved. If these waters are rich in silica, chert is precipitated in these spaces.

An interesting feature of the Kaibab Formation in south-central Utah is small cavities or *vugs* in the limestone that may be lined with a concentric array of quartz or calcite crystals. Some of these vugs contain a thin lining of petroleum residue, and they often emit a fetid odor when broken open. Petroleum is a hydrocarbon compound, meaning it is built of complex molecules composed mostly of hydrogen and carbon atoms. Distinctive compounds found in crude oil indicate one of two organic sources: either chlorophyll from plant material or hemin, the component of blood that gives it its red color. The presence of these compounds provides conclusive evidence that crude oil is the preserved but highly altered product of once-living organisms.

A comparison of the Kaibab Formation at Capitol Reef and the Circle Cliffs with more complete successions that cap the Grand Canyon suggests that much of the formation in southern Utah was stripped away by erosion. Although the Kaibab very likely was thicker and more widespread than present outcrops indicate, its original extent remains speculative (not that this stops geologists!). In the high plateaus of northern Arizona, where the Kaibab caps the imposing walls of the Grand Canyon, geologists are able to subdivide the formation into three informal parts or *members*. The basal "gamma member" consists of sandstone and dolomite with a rich fossil assemblage. The gamma member was laid down as the sea advanced eastward across Utah and Arizona, initially reworking the underlying dune sands, then depositing carbonate as carbonate-forming organisms became abundant. Massive, cliff-forming cherty limestone of the middle "beta member" was deposited while the sea was at its highest level. The uppermost "alpha member" is dominantly red siltstone and sandstone, with evaporite deposits of gypsum that were precipitated as the sea withdrew from the area. In the Circle Cliffs the alpha member is absent and the beta member has been thinned considerably—both as a result of erosion. In addition, wherever it is exposed in south-central Utah, the remaining Kaibab Formation has been cut by canyonlike channels up to 19 m deep (Huntoon and others 1994). Chert pebble conglomerate of the overlying Triassic Moenkopi Formation fills these incisions, indicating that erosion occurred sometime between the Late Permian and the Early Triassic. Chert pebbles probably were locally derived from erosion of the cherty beta member of the Kaibab.

Post-Kaibab erosion was driven by several factors. First, the withdrawal of the sea from the region exposed the Kaibab to relentless scouring by running water. Second, and probably equally important, localized tectonic activity elevated the Circle Cliffs area above the surrounding region, accelerating erosion and funneling detritus into adjacent low-lying areas. This mildly elevated area is called the Emery Uplift; approximately the same area was also uplifted in the Paleozoic, Mesozoic, and even the more recent Cenozoic Era.

References ≈

Baars, D. L. 1962. Permian system of Colorado Plateau. *Bulletin of the American Association of Petroleum Geologists* 46:149–218.

Baars, D. L., and G. M. Stevenson. 1981. Tectonic evolution of the Paradox Basin, Utah and Colorado. In *Geology of the Paradox Basin*, ed. D. L. Weigand, pp. 23–31. Denver: Rocky Mountain Association of Geologists.

Blakey, R. C. 1996. Permian eolian deposits, sequences, and sequence boundaries, Colorado Plateau. In *Paleozoic systems of the Rocky mountain region*, ed. M. W. Longman and M. D. Sonnenfeld, pp. 405–26. Denver: Rocky Mountain Section, Society for Sedimentary Geology.

Blakey, R. C., and R. Knepp. 1989. Pennsylvanian and Permian geology of Arizona. In *Geologic evolution of Arizona*, ed. J. P. Jenney and S. J. Reynolds, pp. 313–47. *Arizona Geological Society Digest* 17.

Chan, M. A. 1989. Erg margin of the Permian White Rim Sandstone, SE Utah. *Sedimentology* 36:235–51.

Condon, S. M. 1997. *Geology of the Pennsylvanian and Permian Cutler Group and Permian Kaibab Limestone in the Paradox basin, southeastern Utah and southwestern Colorado*. Bulletin 2000-P. Washington, D.C.: United States Geological Survey. 46 pp.

Dubiel, R. F., J. E. Huntoon, S. M. Condon, and J. D. Stanesco. 1996. Permian deposystems, paleogeography, and paleoclimate of the Paradox Basin and vicinity. In *Paleozoic systems of the Rocky Mountain region*, ed. M. W. Longman and M. D. Sonnenfeld, pp. 427–44. Denver: Rocky Mountain Section, Society for Sedimentary Geology.

Huntoon, J. E., and M. A. Chan. 1987. Marine origin of paleotopographic relief on eolian White Rim Sandstone (Permian), Elaterite Basin, Utah. *American Association of Petroleum Geologists Bulletin* 71:1035–45.

Huntoon, J. E., R. F. Dubiel, and J. D. Stanesco. 1994. Tectonic influence on development of the Permian-Triassic unconformity and basal Triassic strata, Paradox basin, southeastern Utah. In *Mesozoic systems of the Rocky Mountain region, USA*, ed. M. V. Caputo, J. A. Peterson, and K. J. Franczyk, pp. 109–31. Denver: Rocky Mountain Section, Society for Sedimentary Geology.

Kamola, D. L., and M. A. Chan. 1988. Coastal dune facies, Permian Cutler Formation (White Rim Sandstone), Capitol Reef National Park area, southern Utah. *Sedimentary Geology* 56:341–56.

Kamola, D. L., and J. E. Huntoon. 1994. Changes in the rate of transgression across the Permian White Rim Sandstone, southern Utah. *Journal of Sedimentary Research* B64:202–10.

Kluth, C. F. 1981. Plate tectonics of the Ancestral Rocky Mountains. *Geology* 9:10–15.

McKee, J. W., N. W. Jones, and T. H. Anderson. 1988. Las Delicias basin: a record of late Paleozoic arc volcanism in northeastern Mexico. *Geology* 16:37–40.

Steele, B. A. 1987. *Depositional environments of the White Rim Sandstone Member of the Permian Cutler Formation.* Bulletin 1592. Washington, D.C.: United States Geological Survey. 20 pp.

Ye, H., L. Royden, C. Burchfiel, and M. Schuepback. 1996. Late Paleozoic deformation of interior North America: The greater Ancestral Rocky Mountains. *American Association of Petroleum Geologists Bulletin* 80:1397–1432.

3

Triassic Period

The Triassic Period marks the beginning of the Mesozoic Era and ranges from 245 to 208 m.y.b.p., a span of 37 million years. The Triassic heralds the beginning of the age of dinosaurs, and the remains of these early reptiles have been discovered in Triassic strata throughout the Colorado Plateau. The first land mammals appeared in the Late Triassic. These early mammals were small and rodentlike and remained so throughout the Mesozoic Era.

This period in south-central Utah and most of the Colorado Plateau is represented by two distinctive rock units, the Lower Triassic Moenkopi Formation and the Upper Triassic Chinle Formation. No Middle Triassic rocks are present, signifying a gap between the two formations that represents about 20 million years of missing geologic history. Gaps in the rock record, known as *unconformities*, collectively represent more time than is preserved in the rock record! In general, an unconformity may represent simple nondeposition for that period, or it may mark an erosional event in which strata were present at one time but were eroded before deposition of the overlying unit.

Triassic sedimentary rocks in southern Utah and throughout the Colorado Plateau form the brilliantly colored slopes that apron the imposing cliffs of the overlying Jurassic rocks. The easily erodable nature of these slope-forming rocks is due to their composition, which is in large part shale and siltstone.

The Moenkopi Formation forms brick-red ledgy slopes and often displays conspicuous ripple marks at the tops of the thin beds, indicating a shallow-water origin. The Moenkopi is composed of red siltstone and fine-grained sandstone that record deposition in a variety of arid-climate, low-energy environments. The principal feature of the region during Moenkopi deposition was an arid, windswept shoreline. West of this north-trending shoreline lay a shallow sea; to the east was the flat, monotonous landscape of a fluvial floodplain through which rivers slowly wound their way westward to the shallow seaway that occupied western Utah and Nevada. This floodplain was unbroken all the way to western Colorado, where the mountainous Ancestral Rockies headwaters had been established for some time.

The various shades of red that characterize the sedimentary rocks of the Colorado Plateau region are the result of very small amounts of *hematite*, a red iron oxide mineral that is distributed through the rock. Hematite forms when unstable, iron-bearing minerals break down during weathering and the free iron combines with readily available oxygen. Chemically, hematite is the same as common rust. When iron-bearing fluids flow through unconsolidated sediment, calcium carbonate or silica may be precipitated within the pore spaces between the sediment particles. This cements the loose sediment, transforming it into a sedimentary rock, a process called *cementation* or *lithification*. Tiny amounts of hematite precipitated with the cement are enough to produce vast amounts of red color, resulting in red rocks.

The overlying Chinle Formation is dominated by shale and siltstone that weather to form slopes and isolated badlands. These rocks, some of the most colorful in the region, represent floodplain deposits of an extensive river system that occupied the Colorado Plateau at this time. The colors vary from shades of blue-gray to black and shades of lavender to deep maroon. More resistant sandstone and conglomerate cap the flat-topped mesas and benches and represent river channel deposits. Locally, logs of resistant petrified wood weather out of the sedimentary matrix and litter the slopes and surrounding badlands. Original organic material in the trees has long since been replaced by silica introduced by fluids that flushed through the sediment. Where silica-rich fluids contained traces of iron or copper, the petrified wood is very colorful, with brilliant hues of red, yellow, and purple. Much of the wood in the Chinle is black or gray, reflecting an organic component or simply the natural color of the original wood.

By Late Triassic time the sea that had figured so prominently in Moenkopi geography had receded far to the west and was no longer a factor on the Colorado Plateau. The vast Chinle basin became bounded to the west (present-day eastern California) by volcanic highlands that episodically erupted ash into the air to be carried into the basin by prevailing westerly winds. To the south the basin also was bounded by highlands, perhaps a southern continuation of the western volcanic highlands. The Late Triassic also saw a dramatic shift from the arid conditions of Moenkopi time to a wet subtropical climate. The Colorado Plateau in this period was the site of a vast tropical forest, cut by northwest-flowing rivers and occupied by densely vegetated swamps and wetlands. While terrestrial dinosaurs roamed the floodplains and forests, aquatic reptiles lived in the lakes and swampy areas.

Tectonic Setting ≈

During the earliest Triassic Period the supercontinent of Pangaea was centered around the paleoequator and extended longitudinally from the North to the South Pole (fig. 2.1). Western North America formed the equatorial west coast of the supercontinent, a location that was strongly affected by sea-level variations and climatic patterns.

In a continuation of Late Permian geography, the west margin of Pangaea was initially an ocean, although a volcanic island lay an unknown distance offshore to the west. This *island arc*, as these volcanic islands are called, was separated from mainland North America by a deep marine basin. Shoreline and shallow-marine shelf deposits of the Moenkopi Formation in Utah and Arizona graded westward into a thick succession of marine sedimentary rocks deposited in a westward-deepening seaway. To the east the Late Paleozoic Ancestral Rockies of western Colorado and New Mexico endured and continued to shed sediment westward and influence drainage patterns (fig. 3.1).

Sometime during the Early to Middle Triassic, the western continental margin was disrupted by collision with the outlying volcanic island arc that

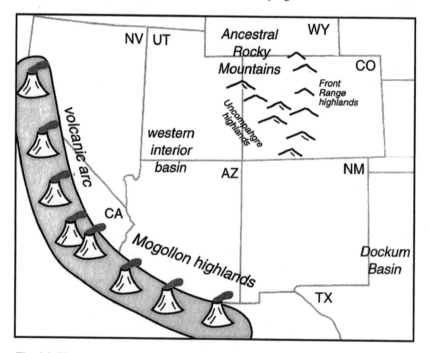

Fig. 3.1. Tectonic features active during the Triassic Period that influenced the paleogeographic evolution of the western interior of North America. After Dubiel 1994.

was rafted in during subduction. This collisional event, called the *Sonoma orogeny*, thrust rocks of the volcanic arc and the intervening marine basin on top of the earlier continental margin. This mountain belt formed the west margin of the Late Triassic Chinle basin and to a certain extent barred the sea from the western interior. Soon after the collision that extinguished the old arc, a new arc became established along the west margin of the continent. This arc covered most of what is now eastern California and some of western Nevada (fig. 3.1). The Chinle basin occupied most of the western interior of North America.

At the same time that the west margin was reorganizing, the Pangaean supercontinent began to break up. Rifts developed along the east and south margins of North America as the breakup initiated. As the South American continent broke away from North America by rotating southwestward, the Gulf of Mexico began to open. It has been proposed that highlands associated with this particular rift margin formed an uplift in southern Arizona and northern Mexico called the Mogollon highlands (fig. 3.1). More recent interpretations, however, suggest that they were simply the southern continuation of the volcanic arc in eastern California that curved southeast into northern Mexico. Regardless of their origin, these highlands were an important sediment source during Triassic sedimentation in northern Arizona and Utah.

Moenkopi Formation ≈

The Lower Triassic Moenkopi Formation in south-central Utah consists of alternating layers of red sandstone and shale that erode into a ledge and slope-type topography. The Moenkopi Formation is extensively exposed in the Circle Cliffs Uplift, Capitol Reef, and the San Rafael Swell, where it forms low brick-red hills broken by rare cliffs. The fine-grained sedimentary rocks that dominate the Moenkopi Formation represent an assortment of low-energy marine and continental environments.

Throughout Moenkopi deposition the sea lay to the west. A north-northeast–trending shoreline extended through southwest Utah and continued northward through north-central Utah. South-central Utah was a vast, featureless coastal plain broken only by northwest-flowing streams that drained distant mountainous areas in western Colorado, New Mexico, and southern Arizona. The very low gradient of the west-sloping coastal plain resulted in major shifts of the shoreline during only minor fluctuations in sea level. A slight sea-level drop or *regression* would cause earlier muddy tidal-flat sediments to be exposed and eventually covered with sand of the west-prograding rivers. Conversely, a rise in sea level or *transgression* would result in sandy river deposits being inundated by the sea and buried by muddy tidal-flat deposits. Four large-scale rises and falls in sea level have been documented in southern Utah during Moenkopi deposition (Blakey and others 1993). Major sea-level changes, coupled with smaller fluctuations, and variations in sediment supply are responsible for the thin-bedded sandstone and shale alternations that characterize the Moenkopi Formation.

An arid climate in southern Utah during the Early Triassic is evidenced by numerous features in the Moenkopi Formation. The evaporite mineral gypsum is common in shallow-marine deposits in southwest Utah and suggests a high evaporation rate. Evaporite minerals such as gypsum and halite are precipitated from mineral-rich waters when evaporation rates exceed the influx of water (for instance, when a basin of sea water gets cut off from the sea). Similarly, gypsiferous siltstone throughout southern Utah suggests an arid sabkha setting. A *sabkha* is a low-relief transition zone between the shallow, tidally influenced shoreline and the purely continental system (eolian or fluvial) that is present landward. Even in an arid climate, a high water table caused by a rise in sea level produces a wet area that only rarely is inundated by the sea. In such a setting high evaporation rates result in the precipitation of evaporite minerals within fine-grained sediment. Sabkhas may also be the site of abundant blue-green algae growth. Blue-green algae are one of the few organisms that can tolerate such extreme conditions. Because competition for such a harsh ecologic niche is rare, algae tend to thrive. Fossil blue-green algae are one of the longest-lived organisms on earth. *Stromatolites*, which are composed of blue-green algae and resemble a head of cabbage, provide our earliest evidence for fossil life in Precambrian rocks and continue to thrive today. As the algae grow, they produce wavy laminae of calcium carbonate. The combination of gypsum, calcium carbonate (produced by algae), and fine-grained sediment that is washed or blown into the sabkha yields a characteristic deposit of wavy-laminated impure sandstone or siltstone. A modern analogy for the Moenkopi sabkhas is the flat, arid region of the southern Persian Gulf. Finally, the presence of eolian and ephemeral stream deposits based on sedimentary structures and textures in some sandstone units strongly supports an arid climate (Blakey and others 1993).

Paleogeography and Stratigraphic Evolution ≈
Following the recession of the Permian sea from central Utah, an interval of subaerial exposure and minor erosion of the Kaibab Limestone ensued. The basal chert pebble conglomerate in the Black Dragon Member in southern

Utah and correlative strata in the Grand Canyon region mark the initiation of Moenkopi deposition (fig. 3.2). Chert pebbles originated from local erosion of the Kaibab Limestone. Like quartz, chert is a mineral that is especially resistant to breakdown by normal weathering processes. In contrast, limestone, which typically is the host rock for chert, is easily dissolved by the slightly acidic composition of natural waters. Because of these extreme differences, residual chert grains are often the only clue to a limestone sediment source.

Finer sand and mud in the Black Dragon Member probably came from the east, brought by sluggish rivers that drained the Uncompahgre highlands of western Colorado. These highlands were the westernmost expression of the much reduced Ancestral Rockies. The age of the earliest sediments shed off the highlands suggests that initial uplift was during the earliest Pennsylvanian Period. By the Triassic, erosion had probably worn the mountains down somewhat. Conglomerate in the Moenkopi in east-central Utah, adjacent to these highlands, however, attests to their continued role as a sediment source of at least moderate relief.

A gradual rise in sea level pushed the northeast-trending shoreline to the southeast, into the Capitol Reef–Escalante Canyons area, covering the basal river deposits. The overlying Sinbad Member records this rise in sea level and consists mostly of limestone deposited in a shallow-marine environment (fig. 3.2). These deposits mark the highest level that the sea reached during the Triassic Period.

After about a million years of marine carbonate deposition, the shoreline abruptly shifted back to the northwest, into western Utah. This rapid shift was caused by a drop in sea level and a simultaneous inpouring of fine sediment from the east. Shale and siltstone overlie the Sinbad Limestone and

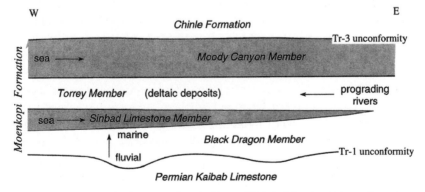

Fig. 3.2. Stratigraphic relations within the Lower Triassic Moenkopi Formation, showing the bounding unconformities and members in the Capitol Reef and Circle Cliffs area, south-central Utah.

mark a change to a broad silt- and mud-dominated tidal flat. The delicate balance of the tidal flat was maintained during a subsequent sea-level rise by a large influx of sediment. West-flowing rivers that drained vast areas of western Colorado, New Mexico, and possibly southern Arizona apparently were rejuvenated at this time. Eventually, the rivers became strong enough to move so much sediment that sandy deltas formed where the larger rivers met the sea. As sea level dropped and the shoreline again receded north-westward, a large delta also shifted northwest into south-central Utah, covering earlier tidal-flat muds (fig. 3.3). Thin deltaic sandstone beds that interfinger with and overlie shallow-marine deposits define the base of the Torrey Member (fig. 3.2), named for the town of Torrey, Utah, west of Capitol Reef. The lobe-shaped delta extended from southeast to south-central Utah and was bounded by shallow-marine embayments to the northeast, along the Utah-Colorado border, and to the south, at the Utah-Arizona border (fig. 3.3). The shallow open sea lay to the northwest.

A reduction of incoming sediment stunted the growth of the delta and caused the uppermost delta sands to be reworked by waves and currents of the adjacent sea. Sand was redistributed laterally, producing thin, wide-spread sheets of sandstone and siltstone along the length of the shoreline. These sheetlike deposits define the top of the Torrey Member. As sea level rose, the delta was drowned and the shoreline advanced eastward. South-central Utah again became the scene of a broad, low-relief shoreline that was greatly influenced by small fluctuations in sea level. This interval is represented by shallow-marine and tidal-flat deposits of the youngest unit of the Moenkopi Formation in south-central Utah, the Moody Canyon Member (fig. 3.2). The wide variety of the rock types in this member reflects subtle changes in sea level and current energy. Shale represents periods of quiet, shallow-water deposition. Sandy siltstone with abundant ripple marks records intermittent current activity, while gypsum indicates high evaporation rates on an arid tidal flat. Interbedded with these relatively low-energy deposits are continuous, ledge-forming sandy siltstone beds that have been interpreted to represent occasional storm events (Blakey 1974).

Latest Moenkopi deposition was followed by an interval of regional erosion throughout southern Utah and northern Arizona that spanned the Middle Triassic, a period of about 20 million years. Erosion resulted in the incision of north- and northwest-trending paleovalleys into the upper Moenkopi that locally are 50 m deep (Blakey and Gubitosa 1983). The unconformable surface has been designated the Tr-3 unconformity. Paleovalleys are filled with conglomerate and sandstone of the basal Shinarump Member of the Late Triassic Chinle Formation.

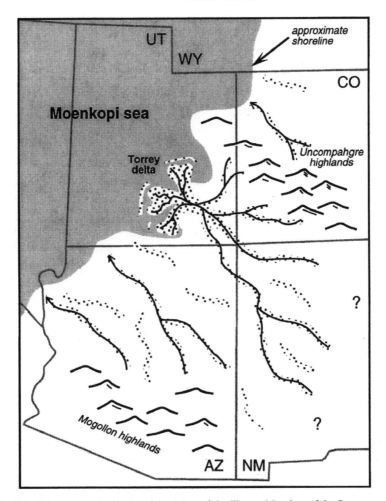

Fig. 3.3. Paleogeography during deposition of the Torrey Member of the Lower Triassic Moenkopi Formation on the Colorado Plateau. After Blakey and others 1993.

Uranium on the Colorado Plateau ≈

Although radioactive minerals had been mined on the Colorado Plateau since about 1905, interest in uranium on the Plateau can be traced back to 1946, when the Atomic Energy Commission (AEC) was created by President Harry Truman. By the late 1940s the AEC, in response to government demand for more weapons-grade uranium, had established sixteen ore-buying stations around the country, twelve of which were located around the Plateau region. Coupled with an increasing number of small uranium discoveries, this confirmed the region's status as the most important uranium district in North America. It also drew the attention of prospectors and speculators from around the nation.

The early 1950s saw the dawn of a true uranium boom in the region. This escalation was driven by generous government incentives to the uranium industry to identify and develop new prospects. This unleashed an army of hardy prospectors across the canyon country, in search of the yellow ore. It appears that these people left no canyon unexplored in their frenzy. Even today remnant tracks can be recognized, often leading to precipitous cliff edges cut by impossibly narrow roads that end for no apparent reason. If one explores further, however, prospect pits in the colorful strata of the Chinle or Morrison Formations can be picked out. These are the rocks that the uranium prospectors sought: these formations hold 85% of the identified uranium deposits on the Colorado Plateau.

United States Geological Survey (USGS) geologists, also supported by the AEC, were one part of the boom on the Colorado Plateau. USGS scientists spent numerous field seasons in the canyons of the Plateau, intensively studying the Chinle and Morrison deposits in an effort to understand their association with uranium. Much of our basic knowledge of the stratigraphy and depositional environments of the strata of the Colorado Plateau comes from these early studies.

Although no definitive understanding of the origin of uranium mineralization was gleaned, several general associations became apparent. Uranium seemed to be concentrated in the lens-shaped sandstone and conglomerate deposits of ancient river channels. The higher porosity of this coarse sediment may have enhanced the passage of uranium-rich fluids through these channel-forms. Channel-forms typically are encased in fine-grained siltstone and shale that would confine the fluids to the coarser sediment, reducing the possibility of dissemination. Additionally, some uranium deposits are associated with altered volcanic deposits such as ash or *volcanogenic sands* (sand grains formed from the erosion of volcanic rocks). It is possible that radioactive minerals originated in volcanic materials and were mobilized by subsurface fluids to be concentrated in the fluvial channel deposits.

Chinle Formation ≈

The Upper Triassic Chinle Formation throughout the Colorado Plateau consists of conglomerate, sandstone, siltstone, shale, and limestone deposited in a spectrum of continental environments, including fluvial-channel and floodplain, lacustrine, and marsh settings. The formation in south-central Utah is more than 500 feet thick and is well exposed in the Circle Cliffs east of the Escalante Canyons and northward into Capitol Reef, west of the Waterpocket Fold. Conglomerate of the basal Shinarump Member forms resistant gray-brown cliffs and, with the remainder of the Chinle Formation, constitutes the Chocolate Cliffs step of the Grand Staircase. Overlying members of the formation form brilliant-hued shale slopes of purple, red, green, gray, and brown. The surface of these colorful slopes has a curious popcornlike texture caused by the clay minerals in the shale. Chinle clays dominantly are bentonite, a clay mineral created through the alteration of volcanic ash. When it is wetted, its crystal structure allows it to absorb water and expand. Upon drying, the clay does not shrink back to its original dimensions, producing the peculiar puffy texture. It is the colorful rocks of the Chinle Formation that define the Painted Desert region and the Petrified Forest of northern Arizona. Chinle slopes are often littered with orange blocks of the overlying Wingate Sandstone, whose vertical unbroken cliffs tower over the varicolored hills. Petrified logs are abundant in the Chinle Formation; because they are so much more resistant to weathering than the fine-grained sediments that encase them, they are often seen lying on the slopes and canyon floors.

Throughout deposition of the Chinle Formation the Colorado Plateau was dominated by a large regional, northwest-flowing river system that probably had an outlet in northwest Nevada. This regional trunk river system was fed by an extensive network of tributaries that drained a large part of western North America. No marine deposits are present in the formation, as the sea apparently had receded from the western interior by this time.

Headwaters of the trunk river were located in the Dockum basin in Texas, Oklahoma, and eastern New Mexico (fig. 3.4). The Dockum basin drained highlands that developed during the late Paleozoic Ouachita orogeny in Texas and Oklahoma. Early interpretations suggested that the Dockum basin was separated from the Chinle basin to the northwest by a drainage divide. Recent detailed studies, however, indicate a connection between the two basins. Detailed comparisons of the chemistry of sand grains from Upper Triassic sandstone in western Nevada and correlative rocks of the Chinle Formation with rocks from the Dockum basin in northwest Texas reveal a shared, chemically distinctive source that can be traced to igneous rocks exposed to erosion during the Ouachita orogeny in southern Oklahoma and Texas (Riggs and others 1996). Furthermore, other studies of fluvial deposits in the Dockum basin indicate west- to northwest-flowing rivers, strongly suggesting that a major throughgoing transcontinental river originated in the Dockum basin and flowed northwest across the Chinle basin into west Nevada, where westernmost exposures of Upper Triassic continental rocks are located.

Tributaries to the trunk river drained several highland sources, funneling large volumes of sediment to the river. The Ancestral Rockies in Colorado maintained relief into Late Triassic time, and a second large northwest-flowing river system drained the north slope of these highlands. This drainage, called the "Eagle paleovalley" by R. F. Dubiel (1992), paralleled the throughgoing regional river system located to the south (fig. 3.4). The Uncompahgre highlands shed sediment into the trunk system through west-flowing tributaries. To the south, in southern Arizona and northern Mexico, volcanic-derived sediment was shed northward into the basin from the Mogollon highlands. These highlands probably were the southeast extension of the volcanic arc that formed the west basin margin. Abundant volcanic material in the form of ash and coarser sediment also was introduced into the basin from the active volcanic highlands to the west. Some ash likely was blown into the basin during eruptions by westerly winds, but the northeast-flowing tributaries contributed coarser sediment to the trunk river.

The Late Triassic climate of the North American western interior is best described as *monsoonal,* a humid tropical environment with seasonal high

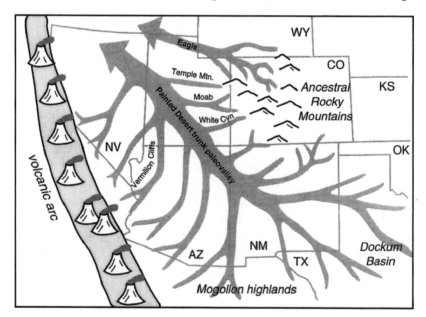

Fig. 3.4. Paleogeography of the western interior of North America during deposition of the Shinarump Member of the Late Triassic Chinle Formation. After Dubiel 1994.

rainfall (Dubiel 1994). A wet environment is implied by the presence of organic-rich lake, marsh, and bog deposits. The fossil record in the Chinle contains abundant, diverse vegetation that indicates a warm, tropical environment. Seasonal rainfall is suggested by features in thick soil zones (*paleosols*, meaning "ancient soils") that developed on the floodplains. The mottled coloring and abundant burrows within the paleosols have been interpreted to indicate a fluctuating water table (Dubiel and others 1987). During seasonal monsoons the water table rose, whereas during the drier periods it would drop but always remained at shallow levels below the surface. The dark purple, lavender, and yellow mottling in the paleosols is caused by variable amounts of iron that were redistributed during alternating oxidizing (dry) and reducing (wet) conditions as the water table fluctuated. Large cylindrical features within the paleosols have been interpreted as crayfish burrows (Hasiotis and Mitchell 1989). Water-table fluctuations caused crayfish to burrow into the soil in an effort to keep their living chambers at the bottom of the burrow below the water table. These trace fossils are abundant in Chinle paleosols.

Sand-dune and playa-lake deposits at the top of the formation suggest that the climate became increasingly arid. A *playa* is a seasonal lake that is common in semiarid to arid environments. The following Jurassic Period marks a return to the arid conditions that prevailed during deposition of the earlier Moenkopi Formation. The Chinle represents a tropical reprieve from the arid climate that otherwise dominated the Triassic and Jurassic Periods.

Paleogeography and Stratigraphic Evolution ≈
The basal Shinarump Member of the Chinle Formation consists of coarse fluvial sandstone and conglomerate that filled incised paleovalleys. Because the Shinarump rivers were confined to paleovalleys, its distribution is irregular but widespread. The Shinarump records deposition in a northwest-flowing regional trunk river and its tributaries that drained areas to the north and south (fig. 3.4). The trunk river originated in the Dockum basin of west Texas, where it was fed by highlands formed during the Late Paleozoic Ouachita orogeny. Tributaries drained the volcanic Mogollon highlands in southern Arizona, the volcanic arc to the west in present-day eastern California, and the Ancestral Rockies in central and western Colorado.

Several discrete paleovalley drainages have been identified and named in southern Utah and northern Arizona. The northwest-trending trunk paleovalley that cut across southern Utah and northern Arizona was named the "Painted Desert paleovalley" by Blakey and Gubitosa (1983).

The "Vermilion Cliffs paleovalley" in northwest Arizona was occupied by a north-flowing tributary that drained the volcanic arc to the southwest (Blakey and Gubitosa 1983; fig. 3.4). The Temple Mountain, Moab, and White Canyon paleovalleys, named for their geographic locations in southern Utah, were west-flowing tributaries that drained the Uncompahgre highlands in western Colorado (Blakey and Gubitosa 1983; Dubiel 1994).

The Shinarump is overlain by finer-grained deposits of the Monitor Butte Member, which reflects the same overall drainage pattern but with a considerable reduction in gradient and energy level. During Monitor Butte deposition nearby volcanic activity dumped huge volumes of ash into the basin. This effectively choked the river channels with sediment, reducing the energy level and damming parts of the system, producing local lakes, swamps, and wetlands. As rivers continued to flow northwest, deltas built into the lakes, eventually filling them with sediment. By the end of Monitor Butte deposition the remnant topography of the original paleovalleys had filled in, smoothing the regional landscape.

The numerous rock types in the Monitor Butte Member represent the diversity of depositional environments during this time. Basal deposits consist of purple mottled paleosol rocks formed during an extended period of weathering and related soil development. Paleosols require long periods to form and reflect intervals of relative stability in which both erosion and sedimentation are minimal. If even moderate erosion occurred, accumulated soil would be washed away; if sedimentation was excessive, soils would quickly be buried and would be thin or poorly developed. This unit contains abundant crayfish burrows (previously discussed), indicating a wet climate.

Overlying rocks consist of black mudstone and shale, with thin limestone beds. Fossils from these rocks include fish scales, fragmented fish bones, and bivalves, suggesting a lacustrine environment (Dubiel 1994). Thin coal beds (≤20 cm) interbedded with lacustrine deposits indicate luxuriant plant growth in a lake-margin marsh or swamp setting. Green clay-rich sandstone and mudstone with large-scale foresets overlie fine-grained lacustrine/wetland deposits. These rocks represent deltas that built into standing water, eventually filling low-lying areas with sediment and establishing an extensive floodplain over which sluggish rivers cut a northwestward path to the sea.

Sandstone, mudstone, and minor conglomerate of the Petrified Forest Member overlie the Monitor Butte Member. These rocks record the recovery of the stagnating northwest-flowing drainage system. The Petrified Forest Member was deposited in a meandering river and floodplain setting pocked with lakes and marshes. The volcanic arc to the west continued to provide ash and coarser sediment to the basin. Overall, the drainage pattern was similar to that of the earlier Shinarump and Monitor Butte Members.

Brown sandstone and conglomerate beds, completely encased in varicolored mudstone, represent discrete river channels. Crossbedding and the geometry of preserved sandbars suggest low-energy, meandering streams. Floodplain mudstones contain limestone nodules that are interpreted to have formed during soil development. In modern settings carbonate nodules have been found to form in soils developed in tropical climates with high seasonal rainfall, a setting that fits well with other climatic evidence from the Chinle Formation. As the Chinle rivers migrated back and forth across the floodplain with time, floodplain sediments were eroded and incorporated into the river channel. Fine clay and silt would be easily swept downstream, while coarser nodules would settle to the channel floor and move only during high-energy flood events.

Thin, discontinuous silty sandstone beds within the mudstone represent deposition on the floodplain during flood events, when sediment-rich floodwater overtopped riverbanks and spilled onto the plain. As overflowing water spread onto the floodplain, energy was immediately reduced, and sand and silt were deposited. Such deposits are called *crevasse splays* and are common in floodplain sediments of the Petrified Forest Member. Local marshes and lakes in low-lying floodplain areas are indicated by black, organic-rich mudstone and the fossils of aquatic phytosaurs and lungfish (Dubiel 1994). Phytosaur fossils are common in the Chinle Formation. These reptiles resembled modern-day alligators and inhabited the waterways of the Chinle basin, including the rivers.

Floodplain mudstone of the Petrified Forest Member grades upward into interbedded limestone and siltstone of the Owl Rock Member. Red and green limestone beds are laterally continuous and signify the development of an extensive lake-marsh system in the basin (fig. 3.5). Limestone beds are thickest in the Four Corners area (up to 3 m) and thin considerably in south-central Utah (< 1.0 m in Capitol Reef), where siltstone dominates. The overall thickness of the Owl Rock Member likewise decreases in all directions from the Four Corners area. Siltstone-dominated strata in Capitol Reef probably represent a lake margin where silt brought in by adjacent streams inhibited the growth of carbonate-producing organisms (fig. 3.5). In order to produce significant amounts of carbonate, most organisms require warm, shallow, clear water. According to Dubiel (1994), a likely modern analogy for the Owl Rock lake-marsh system would be the Everglades ecosystem in southern Florida, where throughflowing, shallow fresh water is disrupted by extensive ponding. Regionally, the Owl Rock Member reflects continued subsidence of the Chinle basin, centered

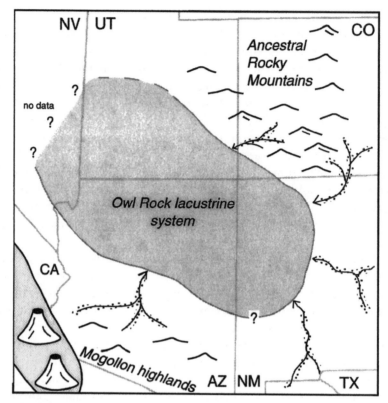

Fig. 3.5. Paleogeography of the Colorado Plateau region during deposition of the Owl Rock Member of the Late Triassic Chinle Formation. After Dubiel 1994.

around the Four Corners area, but with a much reduced sediment supply from surrounding highlands.

The Owl Rock Member is the youngest member of the Chinle Formation in south-central Utah, but not on the Colorado Plateau. The Church Rock and Rock Point Members overlie the Owl Rock to the east, in southeast Utah, northeast Arizona, and central New Mexico. It is possible that these younger members also were present in south-central Utah, but were removed during post-Chinle/pre-Jurassic erosion (J-0 unconformity). Because of this possibility and their climatic implications, these deposits are briefly considered.

The Church Rock and Rock Point Members differ considerably from rocks of the underlying Chinle. Both members consist of red sandstone and siltstone. Some sandstone beds display large-scale crossbedding, a characteristic of eolian sand-dune deposits. The interbedded sandstone and siltstone contain mudcracks, indicating periodic dry conditions. These rocks have been interpreted to represent eolian dune fields traversed by small ephemeral streams (Dubiel and others 1987). One of the more significant aspects of these rocks is the evidence for a major climate change from the tropical monsoonal climate of earlier Chinle deposition to increasingly arid conditions. An arid climate continued well into the Jurassic Period.

References ≈

Baker, S. P., and J. E. Huntoon. 1996. Depositional analysis of the Black Dragon Member of the Triassic Moenkopi Formation, southeastern Utah. In *Geology and resources of the Paradox Basin*, ed. A. C. Huffman, Jr., W. R. Lund, and L. H. Godwin, pp. 173–90. Guidebook 25. Salt Lake City: Utah Geological Association.

Blakey, R. C. 1974. *Stratigraphic and depositional analysis of the Moenkopi Formation, southeastern Utah.* Bulletin 104. Salt Lake City: Utah Geological and Mineral Survey. 81 pp.

Blakey, R. C., E. L. Basham, and M. L. Cook. 1993. Early and Middle Triassic paleo-geography of the Colorado Plateau and vicinity. In *Aspects of Mesozoic geology and paleontology of the Colorado Plateau*, ed. M. Morales, pp. 13–26. Bulletin 59. Flagstaff: Museum of Northern Arizona.

Blakey, R. C., and R. Gubitosa. 1983. Late Triassic paleogeography and depositional history of the Chinle Formation. In *Mesozoic paleogeography of the west-central United States*, ed. M. W. Reynolds and E. D. Dolly, pp. 57–76. Denver: Rocky Mountain Section, Society of Economic Paleontologists and Mineralogists.

Dubiel, R. F. 1992. *Sedimentology and depositional history of the Upper Triassic Chinle Formation in the Uinta-Piceance and Eagle basins, northwestern Colorado and northeastern Utah.* Bulletin 1787-W. Washington, D.C.: United States Geological Survey. 25 pp.

———. 1994. Triassic deposystems, paleogeography, and paleoclimate of the Western Interior. In *Mesozoic systems of the Rocky Mountain region, USA*, ed. M. V. Caputo, J. A. Peterson, and K. J. Franczyk, pp. 133–68. Denver: Rocky Mountain Section, Society for Sedimentary Geology.

Dubiel, R. F., R. H. Blodgett, and T. M. Bown. 1987. Lungfish burrows in the Upper Triassic Chinle and Dolores Formations, Colorado Plateau. *Journal of Sedimentary Petrology* 57:512–21.

Hasiotis, S. T., and C. E. Mitchell. 1989. Lungfish burrows in the Upper Triassic Chinle and Dolores Formations, Colorado Plateau—Discussion: New evidence suggests origin by a burrowing decapod crustacean. *Journal of Sedimentary Petrology* 59:871–75.

Huntoon, J. E., R. F. Dubiel, and J. D. Stanesco. 1994. Tectonic influence on development of the Permian-Triassic unconformity and basal Triassic strata, Paradox basin, southeastern Utah. In *Mesozoic systems of the Rocky Mountain region, USA*, ed. M. V. Caputo, J. A. Peterson, and K. J. Franczyk, pp. 109–31. Denver: Rocky Mountain Section, Society for Sedimentary Geology.

McKee, E. D. 1954. *Stratigraphy and history of the Moenkopi Formation of Triassic Age*. Memoir 61. Geological Society of America. 133 pp.

Riggs, N. R., T. M. Lehman, G. E. Gehrels, and W. R. Dickinson. 1996. Detrital zircon link between the headwaters and terminus of the Upper Triassic Chinle-Dockum paleoriver system. *Science* 273:97–100.

Stewart, J. H., F. G. Poole, and R. W. Wilson. 1972. *Stratigraphy and origin of the Chinle Formation and related Upper Triassic strata in the Colorado Plateau region*. Professional Paper 691. United States Geological Survey. 336 pp.

4

Jurassic Period

Tectonic Setting ≈

The breakup of the Pangaean supercontinent continued into the Jurassic Period. As the south and east margins of North America split from the landmasses of Europe, Africa, and South America, the proto-Atlantic Ocean and Gulf of Mexico gradually widened. The main effect of the breakup on western North America was to disrupt the earlier monsoonal climate patterns. Arid conditions prevailed throughout Jurassic time on the Colorado Plateau.

The southwest margin of the continent was defined by the volcanic arc that initiated during the Middle Triassic. Volcanic-related highlands extended south along the length of eastern California (present-day Sierra Nevada), continuing southeast through western Arizona and into northern Sonora, Mexico (fig. 4.1). Volcanic unrest markedly increased during the Jurassic, probably because of the evolution of an east-dipping subduction zone west of the arc.

A Middle Jurassic tectonic disturbance, the *Nevadan orogeny*, produced a mountainous uplift in central and northern Nevada. Evidence for this event includes folds and thrust faults of Middle Jurassic age, indicating intense compression of the earth's crust during this time. These stresses likely were caused by subduction to the west, as oceanic crust was constantly being stuffed eastward beneath the North American continent. Fold and thrust belt mountains on the continental side of subduction zones are typical of this kind of tectonic setting because of the compressional stresses that are relayed through the crust.

Lower Jurassic ≈

The Jurassic Period saw a return to the arid climate that dominated the Early Triassic of the region. During the Early Jurassic extensive eolian sand seas, called *ergs*, blanketed the Colorado Plateau. Thick accumulations

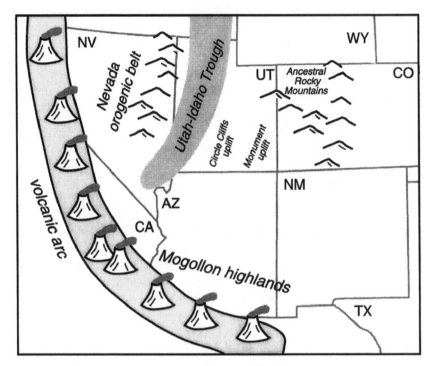

Fig. 4.1. Tectonic features that influenced Jurassic sedimentation and paleo-geographic evolution of the Colorado Plateau. Note that all the features shown probably were not present at any single time. After Peterson 1994.

(more than 2,000 feet) of Jurassic eolian sandstones record millions of years of wind and blowing sand. Large dune fields occasionally were crossed by rivers as they cut the sandy barren expanse on their westward journey to the sea; in some cases these rivers evaporated before reaching it, disappearing into the maze of deep sand and high dunes. Rivers originated in the much reduced Ancestral Rockies of western Colorado.

Lower Jurassic rocks in south-central Utah include, from oldest to youngest, the Wingate Sandstone, Kayenta Formation, and Navajo Sandstone (fig. 4.2). They form a thick sequence of sandstone collectively known as the Glen Canyon Group, named for Glen Canyon, where they form the deep canyon walls of the Colorado River and its tributaries, now mostly lost beneath the waters of Lake Powell.

The Wingate Sandstone in south-central Utah is about 320 feet thick and forms steep, unbroken cliffs. Red-orange Wingate cliffs typically are accented with streaks of blue-black desert varnish formed by a thin coating of manganese oxide. The overlying fluvial Kayenta Formation ranges up to 350 feet thick and is mostly orange to red sandstone with thin mudstone

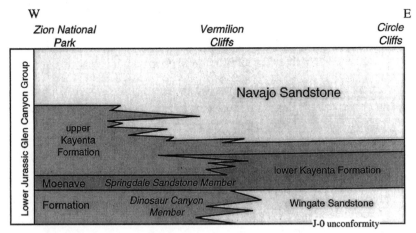

Fig. 4.2. Stratigraphic relations between various eolian deposits (light shading) and fluvial rocks (dark shading) in the Glen Canyon Group, northern Arizona and southern Utah. After Luttrell 1993.

and siltstone layers. Thick sandstone beds form resistant cliffs, whereas thin mudstone and siltstone layers are easily eroded into slopes. The Kayenta weathers into a ledgy steplike interval above the orange Wingate cliffs. The Navajo Sandstone caps the succession and in Capitol Reef consists of about 1,200 feet of white sandstone that has been sculpted into prominent cliffs, domes, and towers. The Navajo is especially spectacular along the crest of the Waterpocket Fold, where it has eroded into a bristling array of rounded domes, turrets, and knife-edged ridges. In fact, early government explorers named Capitol Reef for the resemblance between these high Navajo Sandstone domes and the Capitol dome in Washington, D.C. The Navajo Sandstone forms massive cliffs throughout the Colorado Plateau, including the towering walls of Zion National Park, where it reaches its maximum thickness of more than 2,000 feet.

Wingate Sandstone ≈

Basal Jurassic rocks of the Wingate Sandstone and the related Dinosaur Canyon Member of the Moenave Formation are separated from underlying Triassic rocks by a regional unconformity (J-0 unconformity; fig. 4.2). This surface represents an erosional event that can be recognized throughout the western interior of North America.

The Wingate was deposited in a vast erg that cloaked most of Utah, northeast Arizona, and the western edge of Colorado (fig. 4.3). A northwest-trending zone between Tuba City, Arizona, and Paria, Utah, separates the eolian Wingate to the northeast from coeval ephemeral stream deposits

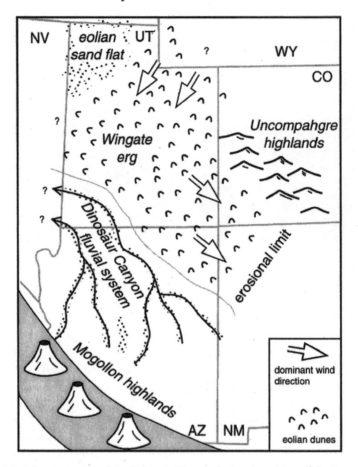

Fig. 4.3. Paleogeography of the Colorado Plateau during deposition of the Lower Jurassic Wingate Sandstone and the Dinosaur Canyon Member of the Moenave Formation. After Peterson 1994.

of the Dinosaur Canyon Member to the southwest. This northwest-trending boundary forms a 25 to 50 km wide transition zone between the two systems.

In the Escalante Canyons and Capitol Reef areas the Wingate Sandstone consists of large-scale crossbedded sandstone and horizontal stratified sandstone that represent large dunes and fringing sand flats, respectively. Thin limestone lenses encased in sandstone reflect the ponding of water between dunes, probably during a rise in the water table. Crossbedding indicates that the wind was blowing to the southeast.

Dinosaur Canyon rivers flowed northwest and defined the southwest margin of the Wingate erg (fig. 4.3). Volcanic-derived river sediment suggests a source in the Mogollon highlands of southern Arizona. It is possible

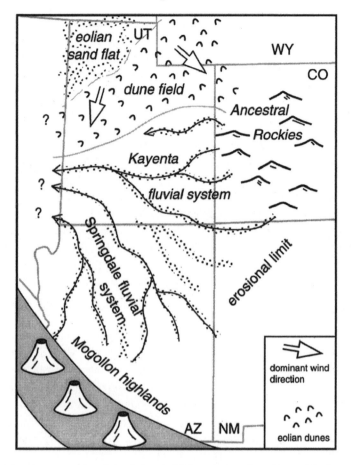

Fig. 4.4. Paleogeography of the Colorado Plateau during deposition of the Lower Jurassic Kayenta Formation and the Springdale Sandstone Member of the Moenave Formation. After Blakey 1994 and Peterson 1994.

that the Ouachita highlands in west Texas continued to contribute sediment as well, although such a source remains speculative.

Kayenta Formation ≈

Nancy Riggs and Ron Blakey (1993) reported a regional unconformity between the Wingate Sandstone and overlying fluvial deposits of the Kayenta Formation. Locally the unconformity is marked by erosion, but in other places it is defined by evidences of simple nondeposition and prolonged exposure to the elements. These include paleosols and traces of tree trunks and roots.

In southern Utah crossbedded sandstones with minor conglomerate and siltstone of the lower Kayenta Formation are the deposits of west-flowing perennial rivers that originated in the Ancestral Rockies (fig. 4.4). The influx of sediment recorded by these fluvial deposits suggests renewed uplift of the Uncompahgre highlands in western Colorado. Volcanic sediment in the correlative Springdale Sandstone in Arizona suggests a source to the south, in the Mogollon highlands. Springdale rivers probably flowed north from southern Arizona to merge with the Kayenta system in southwest Utah. From there the rivers drained into the Utah-Idaho trough, an elongate north-trending depression situated along the Nevada-Utah border (fig. 4.1).

Fluvial deposits of the upper Kayenta Formation in southern Utah and northern Arizona share a complex intertonguing relationship with eolian deposits of the Navajo Sandstone (fig. 4.2). These relations are strikingly similar to those documented in the earlier Wingate–Dinosaur Canyon system (fig. 4.3). The northwest-trending zone of intertonguing passes through northeast Arizona and southwest Utah, separating the upper Kayenta fluvial system to the southwest from the erg to the northeast (fig. 4.2). This zone records the southward advancement of the expanding pile of sand over the north part of the lower Kayenta fluvial plain. Sand-sea expansion pushed the Kayenta fluvial system southward, where it continued to be active, as the erg took over south-central Utah. South of the intertonguing zone the upper Kayenta thickens considerably and is coeval with the basal Navajo Sandstone in south-central Utah. Thus, the gradual southward migration of the Navajo sand sea causes the contact between the Kayenta Formation and the Navajo Sandstone to become younger to the south (fig. 4.2).

Navajo Sandstone ≈

The Navajo Sandstone is the largest preserved eolian system in the geologic record. High cliffs of spectacularly crossbedded white sandstone characterize the Navajo throughout the Colorado Plateau (photo 1). Large-scale crossbeds signify very large, high-relief dunes. In Zion National Park some single crossbed sets indicate dunes with a minimum relief of 50 to 60 feet. Scattered limestone and siltstone beds in the lower part represent ponds that formed in depressions between dunes, probably during times of an elevated water table. Through time, the sand sea expanded over the entire western interior of North America, completely covering the Kayenta rivers to the south. Interbedded Jurassic volcanic rocks and eolian sandstone in southeast California indicate that the eolian system actually lapped onto the volcanic highlands that bounded the basin to the south and west.

Photo 1. Complex crossbedding in the Jurassic Navajo Sandstone in Zion National Park. The sandstone is of eolian origin, and the dip direction of the crossbedding indicates the direction the wind was blowing: dominantly from left to right.

Middle Jurassic ≈

The Middle Jurassic in western North America marks the beginning of an extended interval of mountain-building and related sedimentation. The Nevadan orogeny, appropriately centered in northeast Nevada, initiated this long-lived but sporadic tectonic activity. The development of the Nevadan orogenic belt, which is characterized by folded and thrust-faulted mountains, is attributed to east-directed compressional stresses exerted along the western continental margin. Folding and thrust faulting both shortens and thickens the crust, effectively placing an increased load on the surrounding region. This concentrated load causes increased subsidence or downwarping of the crust, creating a deep basin adjacent to the mountain belt. Such basins are called *foreland basins* because they form in the foreland in front of the fold and thrust belt. It is likely that the accelerated subsidence rate in the Utah-Idaho trough during the Middle Jurassic was driven by crustal thickening in the adjacent orogenic belt (fig. 4.1). The Utah-Idaho trough was centered in northern Utah and southern Idaho, but extended into southwest Utah. Middle Jurassic sedimentary rocks form a rapidly westward-thickening wedge that increases from 430 feet in the San Rafael Swell of central Utah to an astonishing 6,000 feet in the trough in western Utah! The trough gradually shallowed eastward, into south-central Utah. Eastward thinning of strata was disrupted in southeast Utah by the Monument Uplift. Such interrelated zones of uplift and subsidence greatly influenced Middle Jurassic sedimentation patterns on the Colorado Plateau.

In the Middle Jurassic sedimentation on the Colorado Plateau was controlled by a combination of sea level, climate, and tectonic activity. The sea returned to south-central Utah for the first time since the Early Triassic. The Sundance sea entered the region from the north as sea level began a worldwide rise at this time. The climate remained arid, and continental deposition continued to be dominantly eolian.

The interplay among the various factors involved in Middle Jurassic deposition produced complex stratigraphic relations in southern Utah that Ron Blakey of Northern Arizona University and his co-workers have only recently deciphered in detail. Middle Jurassic strata in south-central Utah have been designated the San Rafael Group for the San Rafael Swell, where all the formations are well exposed and easily recognized. The San Rafael Group in south-central Utah, in ascending order, consists of the Temple Cap Sandstone, lower Carmel Formation and correlative Page Sandstone, upper Carmel Formation, Entrada Sandstone, Curtis Formation, and Summerville Formation (fig. 4.5).

Temple Cap Sandstone ≈

The Temple Cap Sandstone is the basal unit in the San Rafael Group and is confined to southwest Utah, in Zion National Park and the St. George area. It is absent in the Escalante Canyons and Capitol Reef areas to the east, and little is known about its original extent. Today it forms a west-thickening wedge of dominantly eolian sandstone. The name "Temple Cap" derives from its position at the top of the massive Navajo Sandstone, which forms many of the "temples" of Zion National Park that were named by early explorers who made forays into the region. The Temple Cap is separated from the underlying Navajo by the J-1 erosion surface, which is marked by a discontinuous layer of chert pebbles. The top of the Temple Cap is defined by the widespread J-2 unconformity (fig. 4.5).

Two members of the Temple Cap Sandstone are recognized in southwest Utah. The basal Sinawava Member consists of red and gray silty sandstone and mudstone of sabkha origin. Gray crossbedded eolian sandstone of the White Throne Member overlies the Sinawava at the top of the

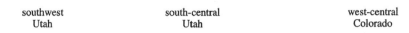

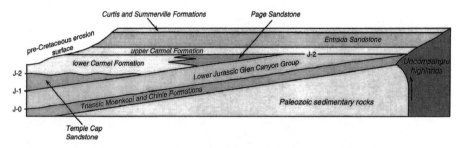

Fig. 4.5. Schematic west-east cross section across southern Utah emphasizing the Middle Jurassic San Rafael Group strata and associated Jurassic unconformities. The San Rafael Group includes all the units between the J-1 unconformity and the top of the Summerville Formation. After Blakey 1994.

canyon walls in the Zion area. To the southwest, near St. George, these upper eolian deposits are replaced by red silty sandstone that is included in the Sinawava. Sabkha deposits in the Temple Cap represent the south margin of a narrow embayment of the Sundance seaway as it inundated the Utah-Idaho trough. Dune deposits of the White Throne Member may previously have stretched much farther east, but were removed by subsequent erosion. The localized preservation of the Temple Cap Sandstone in southwest Utah is likely due to rapid subsidence in the trough at this time. Erosion responsible for the J-2 unconformity causes the formation to wedge out to the east as the J-1 at the bottom and the J-2 at the top converge (fig. 4.5).

Carmel Formation–Page Sandstone ≈
Deposition on the J-2 erosion surface began in southwest Utah with shallow-marine deposits of the Carmel Formation and in south-central Utah with the eolian Page Sandstone. Where the two formations come together in the Paria River area, they display a complex relationship that reflects the interplay between the marine, sabkha, and fluvial systems of the Carmel and the eolian deposits of the Page. The Carmel thickens westward in the Zion area as it approaches the Utah-Idaho trough. The Page Sandstone is confined to south-central Utah and north-central Arizona and is particularly well exposed around the Glen Canyon dam area, where it forms bald red domes and rounded slickrock cliffs of crossbedded sandstone. A short distance to the east the Page pinches out against the Monument Bench, the west side

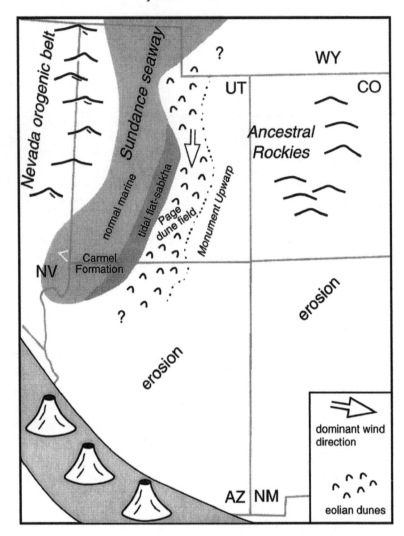

Fig. 4.6. Paleogeography of the Colorado Plateau and surrounding areas during deposition of the Carmel Formation and the Page Sandstone. After Peterson 1994.

of the Monument Upwarp, which apparently was resurrected during the Middle Jurassic.

The narrow seaway that was confined to the Utah-Idaho trough during Temple Cap deposition expanded during this time, extending as far south as northwest Arizona and southern Nevada and eastward into central Utah (fig. 4.6). The thick succession of interbedded fossiliferous limestone and shale in southwest Utah indicates continued rapid subsidence in the

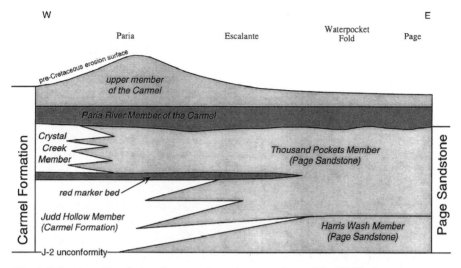

Fig. 4.7. Stratigraphic relations between the various members of the Middle Jurassic Carmel Formation and the Page Sandstone in southern Utah and northern Arizona. After Blakey 1994.

trough. Mountain-building activity in the Nevadan highlands to the west was probably responsible for the high subsidence rate.

Crossbedded sandstone of the Page erg forms a north-trending belt that extends from central Utah southward into north-central Arizona (fig. 4.6). This eolian system was separated from the seaway to the west by a narrow tidal-flat/sabkha complex. During large rises in sea level the coastal complex expanded into the erg. Small sea-level fluctuations, coupled with periodic westward expansion of the erg over the shoreline complex, resulted in the complex intertonguing relations between the Carmel Formation and the Page Sandstone that are so well displayed in the Paria River–Escalante area (fig. 4.7). Farther east, in southeast Utah, the erg was restrained by the Monument Upwarp, a structural high that has been active intermittently through geologic time (fig. 4.6).

In southwest Utah basal fossiliferous marine deposits of the Co-op Creek Member of the Carmel indicate normal marine conditions. The deposits grade eastward into red and gray mudstone and limestone of the Judd Hollow Member, which represent marginal marine tidal-flat and sabkha systems. Correlative rocks in the Escalante–Capitol Reef area make up the Page Sandstone. These eolian deposits are split by a thin tongue of Judd Hollow siltstone that records a marine invasion into the Page erg. This tongue stretches as far east as Waterpocket Fold and serves as the boundary between the lower Harris Wash Member and the upper Thousand Pockets Member of the Page Sandstone (fig. 4.7). As the sea receded, the erg followed, expanding westward over the coastal sabkha. This growth was interrupted by another marine incursion and deposition of the "red marker bed," a thin tongue of red silty sandstone that also reaches to Waterpocket Fold. This was followed by another recession of the sea and the reestablishment of the Page erg over most of its previous range.

After the recession of the sea from the region, a northeast-flowing river system replaced the Judd Hollow coastal complex. The Crystal Creek Member of the Carmel consists dominantly of river deposits and intertongues with the Page Sandstone. Coarse conglomerate at the top of the member contains a chaotic assortment of cobbles and boulders of various volcanic rock types. This unit has been interpreted by M. G. Chapman (1993) as the product of a catastrophic release of floodwater, possibly during the breakout of a crater lake in the volcanic arc some distance to the southwest. Reconstruction of Middle Jurassic paleogeography suggests that if this is indeed the case boulders must have been transported a minimum of 250 km from their source (Chapman 1993). A flood of this magnitude is unheard of in modern times, so that these deposits may represent an extraordinary flood event in the history of the Colorado Plateau.

A short interval of erosion followed Crystal Creek/upper Page deposition, introducing yet another gap into the rock record. In the Glen Canyon and Escalante areas the unconformity is overlain by a transitional member of the Page Sandstone. This transition zone consists of up to 30 feet of sandstone representing eolian dunes and sandsheets and a sabkha environment. Westward, in the Paria area, the unconformity is succeeded by mudstone, gypsum, and carbonate rocks of tidal-flat and sabkha origin. These grade farther west into gypsum and halite deposited in a restricted marine setting. Restriction of the seaway that occupied the Utah-Idaho trough probably was caused by the narrowing of the seaway in northern Utah, where it connected with the open sea to the north. This final marine encroachment onto the Page erg effectively inundated it so that it never recovered. The Sundance seaway in western Utah, however, was also approaching its final days.

The Paria River Member of the Carmel overlies the Page Sandstone in south-central Utah. Along the Paria River these deposits are made up of the detritus of north-flowing streams that carried volcanic boulders up to 3 m in diameter, presumably from the volcanic highlands to the south and southwest. Some workers have pointed to the same distant volcanic source called on for the underlying Crystal Creek Member (Chapman 1993).

Others, however, have inferred a closer source, possibly related to a large explosive eruption of hot ash within the volcanic highlands, but deposited farther north (Riggs and Blakey 1993). If an eruptive blast was focused northward, an ash-rich eruption could disperse hot ash over an extensive area, producing a widespread sheet of densely welded tuff. Depending on the force of the volcanic blast, this conceivably could shift the volcanic source area up to 100 km north of the actual volcano. Since relatively recent uplift and erosion has removed any trace of Jurassic rocks in central Arizona, both possibilities remain unconfirmed. Such is the nature of reconstructing geologic history from an incomplete record.

The volcanic-fed river system was bordered to the west, in the Zion Park area, by a shallow southern arm of the Sundance sea, a small remnant of the sea that once immersed the Utah-Idaho trough. Interbedded gypsum within this interval, between the Paria River and Zion, indicates local sabkha conditions and a continued arid climate. To the north, in the Capitol Reef area, the uppermost Carmel was deposited in a sabkha/tidal-flat system represented by interbeds of red mudstone and siltstone punctuated by thin white gypsum beds.

Entrada Sandstone ≈

As the Sundance sea finally withdrew, pulling north, a growing erg system followed closely on its heels, depositing large-scale crossbedded sandstone of the Entrada Formation over tidal flat mudstones. At its maximum size, eolian sand blanketed much of eastern and southern Utah, as well as large parts of northern Arizona and New Mexico (fig. 4.8). Even the Ancestral Rockies in western Colorado were not immune to the invasion as dunes lapped onto these remnant highlands, covering their foothills and valleys with loose sand.

Tidal flat/sabkha environments still figured prominently in the landscape and covered much of northwest and central Utah. This marginal marine system was associated with the shallow Sundance seaway that earlier had receded northward into central Wyoming. The Entrada erg was bounded to the north and to the west by a muddy coastal plain (fig. 4.8). The sea had vacated the Utah-Idaho trough in western Utah, and the thick sequence of marine limestone and shale that had accumulated over millions of years was buried by an apron of clastic detritus that was dumped off the rising Nevadan highlands.

The Entrada Formation in Capitol Reef consists of structureless white to red-brown sandstone with thin beds of muddy siltstone (photo 2). These strata represent the transition between the erg margin and the tidal flat,

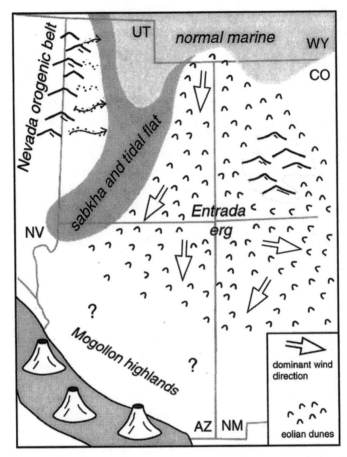

Fig. 4.8. Paleogeography of the Colorado Plateau and adjacent areas during deposition of the Middle Jurassic Entrada Formation and related marine strata. After Peterson 1994 and Riggs and Blakey 1993.

where marine waters reworked eolian sands during high tide or storm events. To the south, eolian crossbedded sandstone becomes an increasingly conspicuous component of the Entrada. Sandstone benches between the lower Escalante River and the Kaiparowits Plateau to the west are capped by red slickrock domes and benches eroded from the crossbedded eolian sandstone. Dance Hall Rock, a natural amphitheater and a prominent landmark in this area, is formed from a large alcove cut into the formation. A north to south transect from Capitol Reef to the lower Escalante canyons clearly illustrates the transition from a marine-influenced erg margin to the erg interior, where marine incursions were relatively rare. After

Photo 2. Pinnacles of Jurassic Entrada Sandstone along the Bullfrog–Notom road in Capitol Reef National Park. Thick, resistant sandstone beds are of eolian origin. Thinly bedded units separating eolian beds are of tidal-flat origin. This succession represents the edge of an eolian dune field that was fringed by a shallow seaway. The shallow seaway sporadically encroached on the eolian dune field, flooding its margin.

the erg had ballooned to its maximum dimensions over the western interior, a period of erosion ensued, followed by an advance of the sea from the north and a contraction of the eolian system.

Curtis and Summerville Formations ≈

Renewed tectonic activity followed Entrada deposition in the western interior and may have provoked the erosion that formed the overlying J-3 unconformity. Coarse sandstone and conglomerate in the Summerville Formation record the first widespread influx of coarse sediment from the Nevadan highlands to the west. The sudden prevalence of coarse sediment suggests that the mountains had risen high above their earlier incarnation. The episodic uplift of these highlands was driven by the eastward migration of the mountain front as compression along the west margin of North America continued. This "ripple effect" through time is typical of folded and thrusted mountain belts. By this time the Nevadan highlands had probably been pushed eastward to the site of the earlier Utah-Idaho trough. Farther into the basin to the east the Circle Cliffs, which previously were the site of sediment accumulation, formed a mild uplift.

Following the brief erosional interval the sea again invaded from the

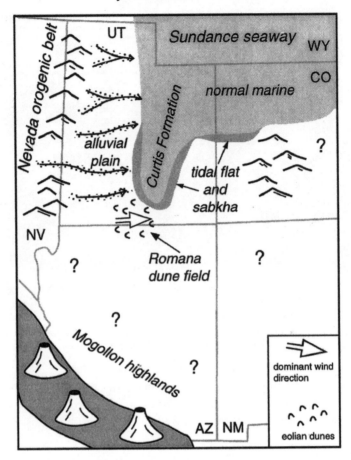

Fig. 4.9. Paleogeography of the Colorado Plateau and adjacent areas during deposition of the Middle Jurassic Curtis Formation and Romana Sandstone. After Peterson 1994.

north, pushing the erg southward. Besides drowning the northern part of the erg, the sea submerged the extensive tidal flats that had furnished sand to the eolian system. The once extensive erg was reduced to a relatively small area centered around the Kaiparowits Plateau. Crossbedded gray sandstone deposited in this dune field forms the Romana Sandstone (fig. 4.9).

Except for the Romana dune field and a fluvial plain that aproned the mountainous region in western Utah, most of the area was a broad mosaic of shallow-marine, tidal-flat, and sabkha environments (fig. 4.9). Marginal marine deposits to the north, in Capitol Reef and the San Rafael Swell, form the Curtis and Summerville Formations. During Curtis deposition a narrow seaway extended southward through Capitol Reef, into the northern Kaiparowits area. Fossils in these rocks point to normal marine conditions. Thinly bedded green mudstone, sandstone, and limestone of the Curtis stand out noticeably from red rocks above and below. The overlying Summerville Formation in all occurrences consists of thin horizontal beds of red and brown mudstone and sandstone and thick white gypsum beds. Preserved ripple marks and mudcracks in these fine-grained rocks indicate initial conditions of a broad tidal flat that periodically was inundated with shallow water. Oscillating currents of the shallow water molded the fine sediment into symmetrical ripples; at other times the sediment was exposed to the dry air, desiccating the thin mud layers and curling them into clay chips to be redistributed by the currents of the next incoming tide. These rocks record a shift to more restricted conditions, as the sea shallowed and became narrower to the north. The upper Summerville in Capitol Reef contains stream-deposited crossbedded sandstone and conglomerate. These streams carried debris from the rejuvenated Nevadan highlands eastward across the western interior of the continent.

The J-5 unconformity developed on top of the Summerville and reflects a period of mild regional erosion, probably as sea level dropped. This surface defines the top of the San Rafael Group and forms the boundary between these Middle Jurassic rocks and the overlying Upper Jurassic Morrison Formation.

Upper Jurassic ≈

Morrison Formation ≈

Arid, desertlike conditions of the western interior during the Early and Middle Jurassic were too harsh for the large dinosaurs that have become associated with the Jurassic Period. During deposition of the Upper Jurassic Morrison Formation, however, the climate became a little wetter, and, based on the number and variety of bones recovered, these large reptiles thrived in the western interior.

The Morrison Formation throughout western North America has yielded hundreds of complete, museum-quality dinosaur specimens. One of the better-known quarries, at Dinosaur National Monument along the northern Utah-Colorado border, is in sandstone of the Morrison Formation that has since been tilted up to nearly vertical. The methods of dinosaur bone recovery can be viewed at the visitor center, where the tilted sandstone beds provide a nicely displayed easel-like setting. Here the paleontologists and other workers painstakingly remove the sediment with tools that may be as fine as dental drills. This particular unit, from which

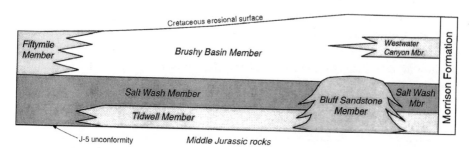

Fig. 4.10. Stratigraphy of the Upper Jurassic Morrison Formation, showing relations with underlying and overlying rocks and between the various members of the formation in southern Utah. After Peterson 1988.

numerous skeletons have been recovered, was deposited on the sandbar of a large river. Apparently, the bodies of dead dinosaurs were washed onto this bar and quickly buried. Rapid burial is required for the preservation of organic material, plant or animal; otherwise, exposure to the air and simple weathering processes would decompose the material, destroying any trace of its existence. Upon burial, silica-rich fluids moving through the sediment replace the organic bone material and fill the pore spaces with silica, producing a very resistant fossilized bone.

The Morrison Formation, like the Chinle Formation, is one of the most economically important producers of uranium ore on the Colorado Plateau. Mineralized zones in the Morrison are mostly confined to discrete lens-shaped fluvial sandstone and conglomerate bodies, a pattern that is reminiscent of the Shinarump ores.

Following a brief interval of erosion Morrison sedimentation began with the return of the sea. The basal Windy Hill Member was deposited in this sea, but is confined to the northern part of the Colorado Plateau, suggesting that the shoreline reached no farther south. Basal Morrison strata in southern Utah are the later Tidwell and Salt Wash Members (fig. 4.10). The Tidwell consists of mudstone and sandstone of a vast mudflat that evolved as the sea withdrew to the north. Rare lens-shaped limestone beds reflect ponding on the mudflat. In the northern Henry Mountains exceptionally thick gypsum beds likely were deposited when marine waters were trapped in an inland evaporative basin, abandoned by the retreating sea.

Southeast Utah was the site of eolian deposition (fig. 4.11). The Bluff

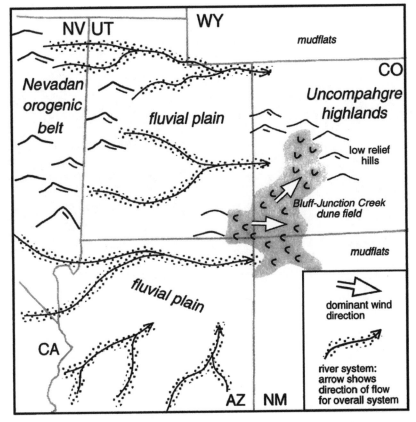

Fig. 4.11. Paleogeography of the Colorado Plateau during Late Jurassic deposition of the Salt Wash Member of the Morrison Formation. High-energy, east-flowing streams originated in the Nevadan orogenic belt to the west and to the south in the Mogollon highlands. Contemporaneous with Salt Wash deposition was the development of a moderate-sized erg in the Four Corners area that produced the Bluff and Junction Creek sandstones. By this time the Ancestral Rockies had been reduced by erosion to low hills. After Peterson 1994.

and Junction Creek Sandstones were deposited in a single, moderate-sized erg that extended from Bluff, Utah, northeastward, well into the Ancestral Rockies of southwest Colorado. The cliff-forming Bluff, a white, large-scale crossbedded sandstone, intertongues with Tidwell mudflat deposits (fig. 4.10). The Bluff Sandstone Member is named for the town of Bluff, where the white sandstone cliffs perch above the town and the nearby San Juan River.

The Salt Wash Member interfingers with and overlies the Tidwell, except in the southeast Kaiparowits basin, where the Salt Wash rests on the J-5

Photo 3. Jurassic Morrison Formation along the Fremont River between Hanksville and Capitol Reef. The lens-shaped sandstone body in the center is a river channel in the Salt Wash Member of the Morrison. Note how it scours deeply into underlying strata. The sandstone is overlain by thinner-bedded fluvial floodplain deposits; these in turn are succeeded by another river channel deposit of sandstone.

erosion surface (fig. 4.10). Sandstone and conglomerate of the Salt Wash are the product of east-flowing, coarse-grained rivers (photo 3), similar to the earlier Summerville rivers that were interrupted by the sea (fig. 4.11). Sediment from these high-energy rivers becomes coarser to the west, toward the Nevadan highland source. Pebbles in the Salt Wash are mostly chert, some containing Late Paleozoic fossils that indicate the erosion of Paleozoic limestone in the highlands (photo 4).

Stream systems like those responsible for deposition of the Salt Wash Member typically originate in mountainous areas, where water is abundant and the gradient of the stream is steepest. Flow is forceful, with a surplus of energy so that erosion, rather than deposition, dominates. In this way mountains are gradually reduced to low hills. This excess energy enables the stream to carry all sizes of particles. The gradient decreases as the stream leaves the mountains, reducing its energy and ability to transport sediment. The stream adjusts by depositing the coarsest sediment. With distance from the mountains the gradient steadily decreases, reducing the stream's ability

Photo 4. Fluvial crossbeds in pebbly sandstone of the Salt Wash Member of the Jurassic Morrison Formation. Crossbed dip direction indicates the direction that Salt Wash rivers flowed, in this case from right to left.

to move increasingly finer sediment. When this concept is viewed on a large, systemwide scale, a gradual downstream fining of sediment becomes evident.

Armed with this knowledge gained from the study of modern rivers, geologists are able to unravel the mysteries of ancient river systems. Because mountain ranges of the distant geologic past may long since have been eroded flat, and even replaced by new ones, the geologist is forced to turn to clues in the debris of the eroded mountains, preserved in the sedimentary record, to reconstruct ancient landscapes. Indeed, this is but one of several lines of evidence used in the reconstruction of the Salt Wash river system.

While Salt Wash rivers coursed eastward, windblown sand continued to accumulate in the Bluff/Junction Creek sand pile (fig. 4.11). Sand was in part blown in from the Salt Wash fluvial system by westerly winds. Evidence for other, smaller dune fields is scattered across the region in rocks of the Tidwell and Salt Wash Members.

Several intrabasinal uplifts became active during Tidwell/Salt Wash deposition. Both units thin and eventually pinch out against the southwest side of the San Rafael Swell. This was the site of the Late Paleozoic Emery Uplift and appears to have been reactivated immediately before or during lower Morrison deposition. Stratigraphic relations in this area provide a clear example of their use in the interpretation of geologic history. The Summerville Formation, which was deposited just before the Morrison, maintains a constant thickness across the uplift, indicating that it was not yet a positive feature. In contrast, the overlying Tidwell and Salt Wash Members thin and pinch out against the uplift and are absent across its inferred crest. Finally, the Brushy Basin Member of the Morrison, which everywhere succeeds the Salt Wash, is present continuously across the area, including the uplift, where it overlies the Summerville Formation. Based on these relations, uplift must have taken place immediately before or during Tidwell and Salt Wash deposition.

During uplift in the San Rafael Swell a similar event was taking place at Black Mesa in northern Arizona. Lower Morrison strata thin and pinch out to the south, against the north flanks of the Black Mesa highland, indicating the presence of a topographic high in this part of the basin.

In southeast Utah fluvial deposits of the Salt Wash also thin, eventually pinching out eastward against sandstones of the Bluff/Junction Creek erg. This pinchout, however, is caused by depositional relief of the erg system rather than by any kind of tectonic activity (figs. 4.10 and 4.11).

The Brushy Basin Member overlies the Salt Wash and correlative strata over most of the Colorado Plateau (fig. 4.10). Red, gray, and green mudstone slopes of the Brushy Basin contrast strongly with the steeper conglomerate and sandstone cliffs of the underlying Salt Wash. Mudstone was deposited on an expansive mudflat dotted with ponds and only rarely cut by east-flowing streams. Sparse units of brown sandstone and conglomerate record the passage of these streams through the mudflat. The pebbles that form the conglomerate are what geologists refer to as the "Christmas tree suite" because they consist mostly of multicolored chert pebbles that resemble brightly colored Christmas tree ornaments.

In the southeast part of the Kaiparowits Plateau strata above the Salt Wash have been defined as the Fiftymile Member of the Morrison Formation (Peterson 1988). As usual, the member is named for its type locality, in this case Fiftymile Point, a promontory at the foot of the Straight Cliffs, which form the northeast ramparts of the Kaiparowits Plateau. In general, a type locality is chosen by the geologist who defines the member (a rigorous procedure in itself) and is the place where the member is best exposed and most easily recognized. The succession of strata at the type locality is called the "type section" and serves as the reference section to which all other localities are compared. All groups, formations, and members must be defined according to a strict set of rules in order to be considered valid.

The Fiftymile Member differs from the correlative Brushy Basin Member in that it is coarser grained and contains more sandstone and conglomerate. Coarser sediment indicates a more proximal position to the source area, which continued to be the Nevadan highlands to the west. Conglomerate composition is dominated by red and green pebbles of the Christmas tree suite. The Fiftymile cannot be traced any farther west because pre-Cretaceous erosion has removed the Morrison Formation in that direction.

The Westwater Canyon Member of the Morrison is confined to the Four Corners area (fig. 4.10). This coarse, crossbedded sandstone interfingers to the north and west with fine-grained strata of the Brushy Basin Member. Westwater Canyon sandstones were deposited in shallow braided river channels that threaded their way eastward across the vast alluvial plain. The energetic, sediment-laden rivers graded, with distance from the source, into finer-grained floodplain deposits of the Brushy Basin. To the west, in western Arizona and Utah, the alluvial plain rose to meet the buttresslike mountain front that provided both coarse detritus and the water with which to transport it.

The close of the Jurassic Period on the Colorado Plateau was a time of erosion rather than deposition. The geometry of the unconformity that developed suggests that uplift occurred along the west margin of the basin, in western Utah and most of Nevada. This erosion surface cuts increasingly older rocks westward, until in the Zion area the Morrison, Romana Sandstone, Entrada, and part of the Carmel Formation are absent. The following Cretaceous Period was a time of dramatic changes in western North America as the earlier Nevadan orogenic belt expanded to a regional scale, with highlands occupying most of Idaho, Nevada, and California and the western portions of Arizona, Utah, and Wyoming. Additionally, a regional seaway returned to the western interior, covering the present-day Colorado Plateau for most of the Cretaceous Period.

References ≈

Bjerrum, C. J., and Dorsey, R. J., 1995. Tectonic controls on deposition of Middle Jurassic strata in a retroarc foreland basin, Utah-Idaho trough, western interior, United States. *Tectonics* 14:962–78.

Blakey, R. C. 1994. Paleogeographic and tectonic controls on some Lower and Middle Jurassic erg deposits, Colorado Plateau. In *Mesozoic systems of the Rocky Mountain region, USA,* ed. M. V. Caputo, J. A. Peterson, and K. J. Franczyk, pp. 273–98. Denver: Rocky Mountain Section, Society for Sedimentary Geology.

Blakey, R. C., K. G. Havholm, and L. S. Jones. 1996. Stratigraphic analysis of eolian interactions with marine and fluvial deposits, Middle Jurassic Page Sandstone and Carmel Formation, Colorado Plateau, U.S.A. *Journal of Sedimentary Research* 66:324–42.

Chapman, M. G. 1993. Catastrophic floods during the Middle Jurassic: Evidence in the upper member and the Crystal Creek Member of the Carmel Formation, southern Utah. In *Mesozoic paleogeography of the western United States II,* ed. G. C. Dunne and K. A. McDougall, pp. 407–16. Los Angeles: Pacific Section, Society of Economic Paleontologists and Mineralogists.

Havholm, K. G., R. C. Blakey, M. Capps, L. S. Jones, D. D. King, and G. Kocurek. 1993. Aeolian genetic stratigraphy: An example from the Middle Jurassic Page Sandstone, Colorado Plateau. In *Aeolian sediments, ancient and modern,* ed. K. Pye and N. Lancaster, pp. 87–107. International Association of Sedimentologists Special Publication 16.

Jones, L. S., and R. C. Blakey. 1993. Erosional remnants and adjacent unconformities along an eolian-marine boundary of the Page Sandstone and Carmel Formation, Middle Jurassic, south-central Utah. *Journal of Sedimentary Petrology* 63:852–59.

Kocurek, G., and R. H. Dott, Jr. 1983. Jurassic paleogeography and paleoclimate of the central and southern Rocky Mountains region. In *Mesozoic paleogeography of the west-central United States,* ed. M. W. Reynolds and E. D. Dolly, pp. 101–16. Denver: Rocky Mountain Section, Society of Economic Paleontologists and Mineralogists.

Lawton, T. F. 1994. Tectonic setting of Mesozoic sedimentary basins, Rocky Mountain region, United States. In *Mesozoic systems of the Rocky Mountain region, USA,* ed. M. V. Caputo, J. A. Peterson, and K. J. Franczyk, pp. 1–25. Denver: Rocky Mountain Section, Society for Sedimentary Geology.

Luttrell, P. R. 1993. Basinwide sedimentation and continuum of paleoflow in an ancient river system: Kayenta Formation (Lower Jurassic), central portion Colorado Plateau. *Sedimentary Geology* 85:411–34.

Peterson, F. 1986. Jurassic paleotectonics in the west-central part of the Colorado Plateau, Utah and Arizona. In *Paleotectonics and sedimentation in the Rocky Mountain region, United States,* ed. J. A. Peterson, pp. 563–96. Memoir 41. Tulsa: American Association of Petroleum Geologists.

———. 1988. Stratigraphy and nomenclature of Middle and Upper Jurassic rocks, western Colorado Plateau, Utah and Arizona. *United States Geological Survey Bulletin* 1633-B:17–56.

———. 1994. Sand dunes, sabkhas, streams, and shallow seas: Jurassic paleogeography in the southern part of the western interior basin. In *Mesozoic systems of the Rocky Mountain region, USA,* ed. M. V. Caputo, J. A. Peterson, and K. J.

Franczyk, pp. 233–72. Denver: Rocky Mountain Section, Society for Sedimentary Geology.

Riggs, N. R., and R. C. Blakey. 1993. Early and Middle Jurassic paleogeography and volcanology of Arizona and adjacent areas. In *Mesozoic paleogeography of the western United States II*, ed. G. C. Dunne and K. A. McDougall, pp. 347–75. Los Angeles: Pacific Section, Society of Economic Paleontologists and Mineralogists.

5

Cretaceous Period

The Cretaceous Period (144 to about 66 m.y.b.p.) saw the completion of the geographic reorganization of western North America that began in the Late Jurassic. In the Four Corners region the Late Cretaceous Western Interior seaway periodically submerged much of eastern Utah and Arizona as well as most of Colorado and New Mexico and reached as far east as central Kansas (fig. 5.1). Furthermore, the seaway stretched northward through Canada and south into Mexico, where it connected with the widening Gulf of Mexico. The western shore of the sea was bordered by an alluvial plain that was cut by large, east-flowing rivers that cascaded from highlands in western Utah and Arizona. Where large, sediment-laden rivers met the shoreline, huge deltas built seaward. The mountainous source area, known as the *Sevier orogenic belt*, was a broad north-trending mountain chain that paralleled the west side of the seaway along its entire extent. Uplift associated with the Sevier orogeny began during the Early Cretaceous and probably was a large-scale escalation of the earlier, more localized Nevadan orogeny. Mountain-building to the west continued until latest Cretaceous time, when the Sevier was succeeded by the Laramide orogeny. The Laramide, while driven by continued compressional deformation of the crust, differed substantially from the Sevier orogeny in structural style (e.g., types of faults, rock types involved in the uplift, etc.) and geographic extent.

Cretaceous rocks on the Colorado Plateau differ dramatically from the pastel-colored shale slopes and fiery orange-red sandstone cliffs that they overlie. Instead, Cretaceous strata are more subdued in color, chiefly shades of gray, brown, and yellow. The colors in these interbedded relics of seas and rivers have been reasonably described as "drab." Thick alternations of easily eroded shale and resistant sandstone cause these rocks to weather into couplets of steep slopes and vertical cliffs that rim extensive plateaus.

The isolated Kaiparowits Plateau is composed solely of Cretaceous strata. Here the 7,000-foot succession has been sculpted into steep cliffs that define the south and east ramparts of the plateau. The great flat-topped plateau is capped by resistant sandstone, sheltering the underlying shale from the scouring forces of erosion. The Kaiparowits is important to geol-

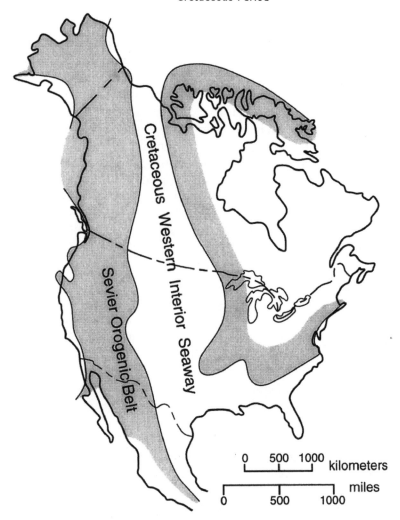

Fig. 5.1. Generalized paleogeographic map of North America during the Cretaceous Period showing the location of the Western Interior seaway, the Sevier orogenic belt, and other emergent areas of the continent. Shaded areas indicate emergent land.

ogists because it contains one of the most complete and best-exposed records of Cretaceous sedimentation in the region.

Another belt of extensively exposed Cretaceous rocks lies along the east flank of the Waterpocket Fold, nestled between Capitol Reef to the west and the lofty Henry Mountains to the east. These rocks can be continuously viewed along the Notom–Bullfrog road, where they offer a phenomenal foreground for the majestic Henry Mountains. The thick sandstone intervals form a series of seemingly impenetrable ridges that are isolated by deep valleys incised into the shale. Cretaceous rocks in the Henry Mountains represent one of the more spectacular landscapes in a region that is defined by sublime panoramas. These strata attain a thickness of only 4,000 feet due to erosion of the upper part of the succession. Large mounds of blue-gray shale exposed along Highway 24 between Caineville and Hanksville appear to have been melted into a mazelike network of badlands, although this landscape is due simply to the forces of water over time. The virtual absence of vegetation coupled with the "melting landscape" style of erosion gives the area a bizarre, moonlike quality.

Because of their easily erodable nature, large areas of Cretaceous rocks have been stripped from the Colorado Plateau. In some places, however, extensive areas have remained intact, notably the Book Cliffs in east-central Utah and adjacent Colorado and Black Mesa, located on the Navajo and Hopi lands of northern Arizona.

The Cretaceous climate of western North America differed considerably from the arid conditions recorded in Jurassic rocks. Several features in Cretaceous rocks on the Colorado Plateau suggest wetter conditions, although there is some argument over *how* wet it was. Significant economic coal deposits are a familiar component of Cretaceous strata in the Four Corners states and have been exploited for more than 100 years. Coal originated from dense mats of vegetation that grew and died in swamps. These plant remnants were then buried and compressed over millions of years to form today's coal deposits. Abundant water is implied, to maintain immense amounts of vegetation.

Additional evidence for a wetter climate, or at least abundant water, lies in the huge amounts of sediment supplied by erosion of the mountains and deposited in large, long-lived deltas along the shoreline. Geologists have long invoked high rainfall to drive the rivers to move such tremendous quantities of sediment. Other geologists, however, have suggested that most of the water originated in the Sevier highlands to the west. These mountains could have acted as a trap for moisture moving east from the Pacific Ocean, producing a rain shadow effect in the low-lying Western Interior basin immediately to the east. This would create a semiarid climate within the basin but provide enough runoff from the mountains to feed the rivers that flowed across it.

A modern analogy would be the present-day Colorado River. The Colorado originates in and is fed by the high Rocky Mountains, where precipitation is abundant, yet it courses through increasingly arid lands on its southward journey to the Gulf of California. Prior to the extensive system of human-made dams, which act as huge sediment traps, the Colorado moved massive amounts of sediment and built an impressive delta at its mouth.

Tectonic Setting ≈

The Sevier orogeny, which remained active throughout the Cretaceous Period, had a profound influence on sedimentation in the foreland basin immediately to the east. The tectonic setting of Cretaceous western North America was very similar to the present-day configuration of western South America. Major features of the Cretaceous North American margin, from west to east, included (1) an east-dipping oceanic plate being subducted beneath the western edge of the North American continent, (2) an active volcanic arc parallel to, and inboard of, the offshore subduction zone, (3) a wide fold and thrust-faulted mountain belt (Sevier orogenic belt) caused by east-directed compression exerted on the continental margin during subduction, and (4) an asymmetric sedimentary basin in which subsidence was driven by the load of the adjacent mountain range (fig. 2.5).

The fold and thrust style of deformation that characterizes the Sevier orogenic belt was the result of intense compression generated by the subduction process. While subduction appears to have been active along this continental margin long before the Cretaceous, research has documented an increased rate of oceanic crust production in the Cretaceous Pacific Ocean. This would certainly have accelerated the rate of subduction, which would in turn increase the intensity of east-directed compression.

Sevier Fold and Thrust Belt ≈

Buckling and breaking during the development of the Sevier highlands mostly was limited to sedimentary rocks and rarely involved deeper-level crystalline rocks. This kind of deformation is referred to as "thin-skinned" because only the upper sedimentary part of the crust appears to have been affected.

Uplift associated with Sevier thrusting initiated in central and western Nevada. Erosion of these early highlands triggered the first pulse of gravel-sized sediment to spread eastward into the downwarping foreland basin. In general, thrust-faulted mountain ranges are a composite of several major thrust faults, with the actively uplifting thrust located immediately adjacent to the basin. This active thrust is called the *frontal thrust* because it forms the front of the actively deforming orogenic belt. As the frontal thrust reaches its limit of deformation, continued compression causes a new frontal thrust to break out basinward or in front of the previous one (fig. 5.2). At this point the earlier thrust becomes inactive, and the new frontal thrust takes over. Because of this sequential development, thrust faults typically are younger toward the foreland basin and older in the direction from which the compression originates. As the thrust belt migrates basinward,

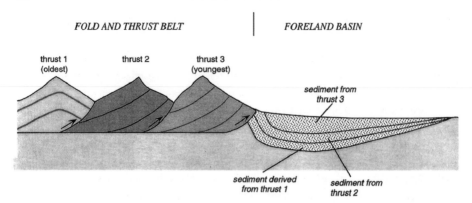

FOLD AND THRUST BELT | *FORELAND BASIN*

thrust 1
(oldest)

thrust 2

thrust 3
(youngest)

sediment from
thrust 3

sediment derived
from thrust 1

sediment from
thrust 2

Fig. 5.2. Diagram showing typical relations between various thrust faults and foreland basin deposits. Note that thrusts become younger to the right, toward the foreland. Additionally, while thrust 1 was contributing sediment to the basin, thrusts 2 and 3 did not yet exist. Subsequently, when thrust 2 formed in front of thrust 1, thrust 1 became inactive and was blocked from contributing sediment to the basin. Similarly, thrust 3 did not exist at that time.

the adjacent basin margin also is deformed with each new thrust. In this way the Sevier thrust belt slowly migrated eastward, inching its way well into western Utah by the end of the Cretaceous Period.

With the formation and uplift of each new thrust in the Sevier belt, earlier sediments in the proximal part of the basin also were uplifted, eroded, and recycled as sediment along the new basin margin. In some instances the new frontal thrust would actually ride over the detritus of the previous thrust, burying and protecting it from erosion. These deposits and those preserved farther east in the basin provide us with many of our clues for the timing and sequence of thrusting. In the case of the Sevier orogeny in Utah we are fortunate that each thrust uplifted a unique and recognizable group of rocks to be eroded and dispersed eastward across the foreland. Through the careful study of the composition of sand and gravel shed off each thrust we see that specific compositions are restricted to certain layers in the basin. By becoming familiar with the rock types in each thrust we can reconstruct the order of thrusting with some accuracy. For instance, the earliest deposits in the basin contain pebbles of an easily recognizable Precambrian-age sandstone. By matching these pebbles (and others) with the thrust system that contained Precambrian sandstone we learn that this thrust was among the earliest to be active in the mountain belt. Furthermore, if the age of these basin deposits can be determined through fossils or some other method, the actual age of movement on that thrust can also be determined.

Western Interior Foreland Basin ≈

The Cretaceous foreland basin that occupied the Colorado Plateau was broadly asymmetric, with individual depositional units dramatically thickening westward, against the Sevier thrust front, and gradually thinning eastward across eastern Utah, Colorado, and western Kansas. Asymmetric basin subsidence was driven by thickening of the crust in the Sevier belt, as massive slabs of layered sedimentary rock were thrust up and over their eastern counterparts. Rapid subsidence along the west margin was a direct response to this crustal thickening. Continuous uplift and erosion of the Sevier belt also provided abundant sediment to the basin, strongly influencing facies patterns seen today.

A spectrum of depositional environments is represented by Cretaceous rocks in the Western Interior basin. In a regional sense, sediments become finer-grained to the east, away from the Sevier mountain front. This fining also corresponds to a gross change in depositional setting. Continental environments occupied the west basin margin. These systems graded eastward, from coarse-grained alluvial fans to rivers. Extensive long-lived coal swamps were common features of low-lying floodplains in the distal, low-energy reaches of these rivers. Where they emptied into the sea, deltas were constructed. Marine environments present in the central and east parts of the basin represent a variety of shoreline, shallow-marine, and deep-sea environments. Like their continental counterparts, marine deposits also decreased in grain size to the east as the sea deepened in that direction.

Cretaceous rocks in southern Utah record numerous shifts between continental and marine environments. Sea-level fluctuations were one important cause of these changes. Deltaic and shoreline environments in south-central Utah were especially sensitive to these fluctuations. When sea level rose even a small amount, low-lying deltas were drowned and the shoreline pushed west. If sea level dropped, rejuvenated rivers cut through deltaic sediments to form new deltas farther east. In such instances coal swamps would become established on older delta deposits, much like the present swamps landward of the modern Mississippi River delta. Tectonic activity in the orogenic belt to the west was another important factor, controlling both the sediment supply and subsidence rates. Thus, the mosaic of deposits seen in the Kaiparowits Plateau and Henry Mountains area mirrors the complex interplay of sea level, sediment supply, and subsidence.

Cedar Mountain Formation ≈

The Lower Cretaceous Cedar Mountain Formation represents the introductory chapter in the Cretaceous story of southern Utah. The Cedar Mountain unconformably overlies the Jurassic Morrison Formation; while it appears that a considerable amount of unrecorded time is represented by this contact, the dearth of datable fossils in the basal Cedar Mountain makes this unconformity speculative and somewhat controversial. The formation is divided into two parts, the basal Buckhorn Conglomerate Member and an overlying unit of gray, green, and purple mudstone. The unnamed upper unit differs appreciably from the drab-colored rocks that make up the rest of the Cretaceous succession. In fact, Cedar Mountain mudstones share a greater resemblance to the colorful strata of the underlying Morrison Formation, a feature that only fuels the controversy over the supposed unconformable contact between the two formations.

Although the Cedar Mountain locally reaches more than 150 feet thick, it is much thinner or absent over much of southern Utah. The formation is continuously exposed in northern Capitol Reef, along the eastern ramps of the Waterpocket Fold. Here the Buckhorn forms a resistant ridge, while the overlying mudstones are carved into valleys. The formation extends northward into the San Rafael country, where its namesake Cedar Mountain is located. The Cedar Mountain Formation is absent in southern Capitol Reef and the Kaiparowits Plateau, although it has been reported around the town of Escalante.

The basal Buckhorn Conglomerate Member consists of conglomerate and coarse sandstone deposited in swift, east-flowing rivers that drained rising highlands to the west. In contrast, the upper part of the Cedar Mountain Formation is dominated by colorful mudstone with only sparse lenses of sandstone and conglomerate. These upper strata represent floodplain mud and silt, occasionally cut by river channels that eventually filled with sand and gravel. Abundant plant fossils, including delicate leaf impressions, indicate a quiet, vegetated floodplain. Thin, discontinuous limestone beds sandwiched within mudstone units probably accumulated in low-lying, poorly drained areas of the floodplain that filled with standing water.

Taken by itself, the Cedar Mountain Formation in southern Utah is an unimpressive reflection of earliest uplift in the Sevier belt. However, if Lower Cretaceous strata are considered regionally, a striking picture of foreland basin development emerges. Investigations by geologist Peter Schwans on equivalent strata in west-central Utah, which he named the Pigeon Creek Formation, revealed an astonishing 3,000 feet of conglomerate, sandstone, and mudstone. This thick succession was deposited immedi-

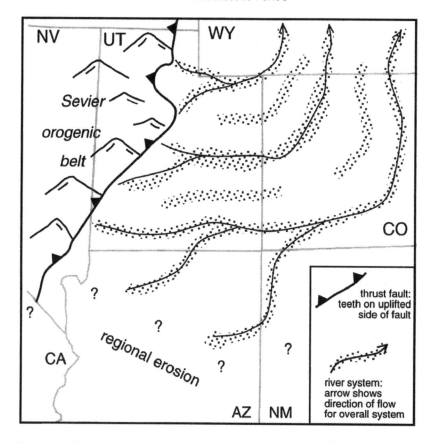

Fig. 5.3. Paleogeography of the Colorado Plateau during Early Cretaceous Cedar Mountain/Burro Canyon deposition. Note the location and flow direction of the river systems and the highlands of the Sevier orogenic belt that fed the rivers. The sea at this time lay to the northeast, in present-day Canada.

ately east of the advancing orogenic belt, where rapid downwarping was driven by the thrust load to the west.

Cedar Mountain rivers traversed eastward from the Sevier highlands, cutting the wide, muddy floodplain in Utah and western Colorado with ribbons of sand and gravel (fig. 5.3). Upon reaching present-day Colorado, rivers turned northward to drain into the incipient Western Interior seaway, which had not yet expanded into the area.

Deposition of the Lower Cretaceous Cedar Mountain Formation was followed by a two- to three-million-year interval of erosion that spanned the Lower/Upper Cretaceous boundary. The present uneven, sporadic distribution of the Cedar Mountain Formation is attributed to local removal by erosion during this time. The incision of valleys through the

Cedar Mountain Formation and locally deep into underlying strata set the stage for the subsequent deposition of the Dakota Sandstone.

Stratigraphic Problems of the Lower Cretaceous ≈

The Cedar Mountain and related formations are one of several rock units on the Colorado Plateau that, rather confusingly, assume a different name with a change in location. Upon traversing eastward across an imaginary north-south line that corresponds with the trace of the Colorado River, the Cedar Mountain Formation becomes the Burro Canyon Formation. While the change from Cedar Mountain to Pigeon Creek Formation to the west is justified by a dramatic increase in thickness and a change in rock type, no such changes occur across the Colorado River. In fact, the Burro Canyon Formation is identical to the Cedar Mountain, and the name change is defined only by a change in geographic setting.

Confusing, seemingly unexplainable name changes in a single rock unit can often be blamed on a lack of communication among the pioneering geologists who established the names. Geologists working on the same strata, but in different areas, commonly named rock units in an effort to establish some order to the succession. Problems arose and continue to plague us when some units ended up with several names. In some cases these names became entrenched through use, evolving into a specific and familiar body of rock to geologists. These scientists generally are a hard-headed lot and strongly resist name changes—even if the purpose is to set the record straight. While geologists may understand the true relations, problems surface when nongeologists attempt to decipher the relations between units. For example, a student of the stratigraphy of the Colorado Plateau may mistakenly think that the Cedar Mountain Formation in Capitol Reef is different from the Burro Canyon Formation in Arches National Park when, in fact, they are the same unit. Sometimes names are changed to reflect stratigraphic relations more accurately, but these changes often occur at a pace that rivals geologic time.

Dakota Sandstone ≈

Mid-Cretaceous erosion was closely followed by deposition of the Late Cretaceous Dakota Sandstone, a widespread unit that was deposited throughout Utah and Colorado and much of northern New Mexico and Arizona. In the Henry Mountains and the Kaiparowits Plateau the Dakota is up to 150 feet thick, although locally it is much thinner. The Dakota forms an amazingly consistent vertical sequence, considering its wide ex-

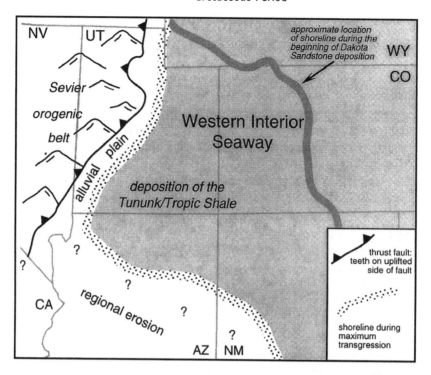

Fig. 5.4. Paleogeography of the Colorado Plateau during the sea-level highstand of the Greenhorn cycle and Tunumk/Tropic Shale deposition. The location of the shoreline during initial Dakota Sandstone deposition is included for reference to show the extent of shoreline migration. After Roberts and Kirschbaum 1995.

tent. The basal Dakota consists of crossbedded sandstone and conglomerate that fill erosional hollows and valleys that were cut as deep as 90 feet into older rocks. Coarse basal deposits grade upward into fine sandstone and siltstone that are rich in black carbonaceous material that includes plant remains and thin coal seams. The uppermost Dakota consists of sandstone with oyster shell horizons and grades upward into black, organic-rich shale.

Dakota deposition began as sea level rose and the shoreline crept westward. This sea-level rise eventually produced a classic transgressive stratigraphic sequence. Basal fluvial sandstone and conglomerate filled the incised valleys. Like earlier Cedar Mountain rivers, Dakota rivers were fed sediment by the rising Sevier highlands to the west (fig. 5.4). These rivers carried sand and gravel eastward to meet the advancing Western Interior sea, whose shoreline initially lay in central Colorado.

As the shoreline inched westward, the lower reaches of the rivers became sluggish. This decrease in energy resulted in deposition of increasingly finer sediment and the accumulation of organic debris such as plant remains.

Local thin coal seams and carbonaceous sediment suggest marshy, swamp-like conditions along the river-shoreline interface.

The shoreline migrated over carbon-rich sand and mud as sea level continued to climb. Shallow-marine sandstone of the upper Dakota displays a multitude of sedimentary structures that variously suggest incoming and outgoing tidal currents, deposition on offshore sandbars, and normal wave activity. Fossil shells and the feeding and dwelling traces of fossil organisms are significant features in these rocks. While geologists may not know exactly what organism made the trace fossils, traces can be correlated with different water depths and/or specific subenvironments of the sea. Oysters and shells of the fossil *Gryphaea* (a clamlike shell that was christened "devil's toenail" by early explorers for its resemblance to a twisted and gnarled toenail) are especially common in the upper marine Dakota along the Waterpocket Fold. Oyster Shell Reef, shown on most topographic maps of Capitol Reef, is located along the Notom–Bullfrog road east of Muley Twist Canyon. This ridge is named for a resistant spine of upper Dakota Sandstone that is packed with oyster shells. These fossils and trace fossils alike indicate a shallow-marine environment with water of normal marine salinity. Through time the sea deepened, and sand was trapped along the new shoreline farther west. Offshore areas only received the clay-sized sediment that was easily winnowed seaward by weakening currents, eventually settling to the deep ocean floor. With time the Dakota was slowly but steadily blanketed by a thick layer of dank black ooze that marks the onset of deep-sea conditions.

The Dakota Sandstone and overlying shale record the oldest of four major transgressive-regressive sea-level cycles recognized in Upper Cretaceous rocks of southern Utah. These sea-level changes were slow, even by geological standards. Not only does the Dakota record the slow westward migration of the Western Interior sea, but the unit itself transgresses time. In other words, as sea level rose and the shoreline advanced, marine sandstone that declares the passage of the shoreline becomes progressively younger to the west. Age-diagnostic fossils in the upper Dakota along the Utah-Colorado border indicate initial marine conditions at about 92.5 million years ago, whereas the same unit to the west, in Price, Utah, is about 91.5 million years old. This translates into a migration rate of 90 miles/m.y. or 0.5 feet/year, which certainly does not imply a steady, continuous advancement. In reality the migration was fitful, at times advancing quite rapidly, at other times remaining in one position for a time or even receding for a short time. While the sea overall marched westward over the fluvial floodplain, this occurred through numerous short-term oscillations.

Tropic Shale and Tununk Member of the Mancos Shale ≈

Blue-gray shale that blankets the Dakota Sandstone throughout southern Utah goes by different names depending on where you are (fig. 5.5). In this case, however, there are reasons for the proliferation of names; the main reason involves facies changes in these rocks from west to east. In the Kaiparowits Plateau and Bryce Canyon areas to the west the unit is called the Tropic Shale and is considered a formation. In contrast, the same shale in the Henry Mountains, 25 miles to the east, is the Tununk Shale Member of the Mancos Shale. Farther east, in the Book Cliffs and Arches National Park, it is not easily divisible into members and is simply the lower Mancos Shale.

In the Bryce Canyon, Kaiparowits, and Henry Mountain areas the unit consists of about 700 feet of shale, with only rare, thin sandstone beds. Regardless of what we call it, the shale was deposited in the deeper reaches of the seaway, in quiet water far from the shore. While sand mostly was confined to the shoreline and nearshore area to the west, this monotonously thick pile of black ooze accumulated offshore, under a slow but constant rain of mud. The uniformity of the thick shale sequence is only rarely broken

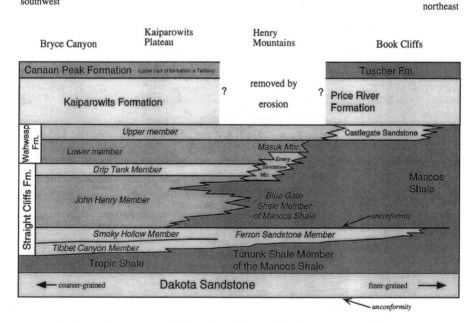

Fig. 5.5. Diagrammatic cross section showing the stratigraphic relations between Upper Cretaceous rock units in southern Utah. After Peterson and others 1980.

by thin beds of sandstone. Sand likely was introduced into the deep sea only during violent storms that generated enough wave activity to sweep sand out into deeper water.

The *highstand* or sea-level maximum of this rise and fall cycle occurred sometime during shale deposition. At its climax the Late Cretaceous sea submerged most of the present-day Colorado Plateau. As deep marine muds blanketed southern Utah, a narrow strip of emergent land in far western Utah partitioned the shoreline from the neighboring mountain front to the west (fig. 5.4). Sand and gravel deposited on this emergent plain by east-flowing rivers form part of the Iron Springs Formation in southwest Utah. During the sea-level highstand the normal transition of a river system from steep, turbulent river to placid meandering stream was truncated by the westward encroachment of the sea. Because the normally broad floodplain was shortened, energetic high-gradient rivers dumped their coarse sand and gravel load directly into the sea.

The sedimentary record from the Late Cretaceous sea shows four large-scale sea-level rise and fall cycles. The recognition of these cycles has been important in identifying the regional effects of sea-level changes on sedimentation. For instance, pre-Dakota erosion and valley development was related to a sea-level minimum or *lowstand*. Subsequent deposition of the Dakota within the confines of these valleys was triggered by the slow rise of the sea. The inundation of the valleys and surrounding areas by the rising sea can be seen in the Dakota as coarse stream-laid sandstone and conglomerate pass upward into fine-grained sandstone with marine fossil shells. Overlying black marine shale of the Tropic/Tununk Shale testifies to much deeper marine conditions as the sea continued to rise. The Dakota Sandstone and overlying shale form the transgressive part of the *Greenhorn cycle*, the first of these late Cretaceous cycles.

Lower Straight Cliffs Formation and Ferron Sandstone Member of the Mancos Shale ≈

Following the Greenhorn highstand sea level dropped, and the alluvial plain expanded eastward. This regression was simultaneous with renewed uplift in the Sevier highlands as the steep mountain front also inched eastward. Rejuvenated rivers distributed coarse sediment across the broad alluvial plain, and deep sea mud was gradually overcome by sand.

Sandstone deposited during the regressive phase of the Greenhorn cycle undergoes name changes in a west to east direction. Again, the different names are related to facies changes between the Kaiparowits Plateau and the Henry Mountains. In the Kaiparowits these sandstones include the basal Tibbet Canyon and overlying Smoky Hollow Members of the Straight Cliffs Formation (fig. 5.5). The formation is named for its spectacularly continuous exposure along the Straight Cliffs, which define the abrupt eastern boundary of the grand, tablelike Kaiparowits Plateau. The Straight Cliffs Formation also includes the John Henry and the Drip Tank Members in its upper part, but these are separated from the underlying Tibbet Canyon and Smoky Hollow Members by an erosional unconformity and are not part of the Greenhorn cycle.

As we move east to the Henry Mountains, the Ferron Sandstone Member of the Mancos Shale is the counterpart to the lower members of the Straight Cliffs Formation (fig. 5.5). The Ferron is an east-thinning tongue of sandstone that splits the thick Mancos succession into the lower Tununk Shale and the overlying Blue Gate Shale Members. The unconformity that caps the Greenhorn cycle to the west likewise extends across the top of the Ferron Sandstone.

In far western Utah coarse debris of the Iron Springs Formation continued to be deposited along the flanks of the advancing Sevier mountain front. Raging rivers that drained these highlands were mostly unaffected by the eastward retreat of the sea. Coarse sand and gravel graded eastward into shoreline and nearshore marine deposits of the basal Straight Cliffs Formation. In the Kaiparowits area fine-grained sandstone of the Tibbet Canyon Member reflects a transition from deep-sea to a shallow shoreline setting as the sea receded from western Utah (fig. 5.5). Evidence for a shallow-marine environment can be seen in abundant fossil shells, including the common Late Cretaceous clamlike bivalve *Inoceramus*, and fossil shark teeth.

Coal and carbon-rich mudstone of the Smoky Hollow Member overlie marine sandstone of the Tibbet Canyon. Where present, coal beds may reach up to four feet thick, but most are thinner. Sandstone, mudstone, and conglomerate overlie carbonaceous strata and form an upward-coarsening sequence that initiates with fine sandstone and mudstone and grades into coarse sandstone and conglomerate. The top of the Smoky Hollow is truncated by an erosional unconformity that can be traced throughout south-central Utah (fig. 5.5).

The Smoky Hollow Member is named for Smoky Hollow, a wide, deep canyon that drains southward off the Kaiparowits Plateau. Here the lower fine-grained beds form a series of slope and ledges, while upper coarse strata form a series of broken cliffs. Along the Straight Cliffs, which overlook the

Escalante drainage to the east, both the Tibbet Canyon and Smoky Hollow Members are concealed beneath talus and debris created by the constant disintegration of overlying Cretaceous rocks.

The Smoky Hollow reflects the continued recession of the sea and the final scene of the Greenhorn sea-level cycle. Coal marks the passage of the sea from the immediate area and the evolution of a poorly drained, thickly vegetated swamp. This coastal swampland likely provided abundant food for large plant-eating dinosaurs, as well as cover from the carnivorous dinosaurs that preyed on them.

As we move east into the Henry Mountains area, the lower part of the Ferron Sandstone Member consists of marine sandstone and mudstone much like its counterpart Tibbet Canyon Member in the Kaiparowits (fig. 5.5). Fossils, trace fossils, and abundant ripple marks indicate shallow-marine conditions that have been interpreted as a barrier bar setting. *Barrier bars* are large, shallow offshore sandbars that parallel the shoreline, protecting it from the pounding wave action of the open sea.

The Tibbet Canyon/Smoky Hollow sequence and the Ferron Sandstone both record the slow eastward recession of the sea. The sequence tracks this sea-level drop as underlying deep-marine shale grades up into shallow-marine sandstone. Continued withdrawal resulted in the eastward advancement of low-energy rivers. The low-lying coastal plain became engulfed with tangled masses of vegetation that eventually produced thick coal deposits. On the Kaiparowits Plateau conglomerate that caps the Smoky Hollow heralds the *progradation* or advancement of coarse-grained, high-energy rivers over the swampy marshlands. As the sea withdrew still further, erosion took over and resulted in an unconformity at the top of the sequence that can be recognized throughout the Four Corners region (fig. 5.5). This unconformity marks the top of the Greenhorn cycle and probably represents several million years of unrecorded time.

The regressive sequence recorded by the lower Straight Cliffs Formation and Ferron Sandstone is a mirror image of the preceding transgressive Dakota/Tununk–Tropic Shale succession (fig. 5.5). Collectively this transgressive-regressive sequence constitutes the Greenhorn cycle. The lower and upper boundaries of the sequence are defined by erosional unconformities that developed during sea level lowstands. The middle of the cycle or "turning around point" for the sea occurs somewhere within the thick Tununk–Tropic Shale sequence.

Upper Straight Cliffs Formation and Blue Gate Shale and Emery Sandstone Members of the Mancos Shale ≈

On the Kaiparowits Plateau the John Henry Member lies on the regional erosion surface that caps the Greenhorn cycle (fig. 5.5). The John Henry varies from 660 to about 1,100 feet thick and, like most of the Cretaceous rocks in the region, has been carved into a series of cliffs and slopes.

Along the southwestern margin and the central part of the Kaiparowits tableland sandstone, mudstone, and coal were deposited on a broad coastal plain. Sand was distributed across the plain by lazy rivers that came from the Sevier highlands. Over millions of years mud collected on the adjacent floodplains, which were periodically replenished with fresh sediment as rivers grew too big for their channels. Black carbonaceous debris in these sediments suggests that vegetation thrived in this humid, low-lying area. Coal, which formed in junglelike coastal swamps, is especially abundant and occurs in several thick beds and numerous thinner ones.

The John Henry Member contains coal in significant economic amounts. While this economic potential has been apparent for some time, the isolation of the Kaiparowits Plateau (or southern Utah for that matter) has so far left this resource unexploited. In the last decade these coals have been the subject of intense study by the U.S. Geological Survey and several coal companies that were intent on mining it. It is the John Henry coal beds, in part, that made the designation of the Grand Staircase–Escalante Canyons National Monument so controversial. When the Kaiparowits was named as part of this expansive monument, the Dutch-owned Andalex Resources was moving forward with plans to mine the coal. These plans were bitterly opposed by environmentalists throughout the world, who saw the Kaiparowits as one of the largest undeveloped tracts of wildland in the lower forty-eight states. Even a few of the citizens of southern Utah were uncharacteristically opposed to the scale of the mining operation. This opposition stemmed mostly from the proposed heavy truck traffic through rural areas that would be required to move the coal to the West Coast, where it was to be exported for use in foreign power plants. Mining claims by Andalex Resources are currently being negotiated with the U.S. government.

The thickest coal bed in the area is an impressive 29 feet thick, although this bed is exceptional and probably thins quite a bit laterally. Most of the coal beds in the John Henry are less than 4 feet thick. Coal is thickest in the central part of the Kaiparowits and thins in all directions. These coal beds represent surprisingly stable and long-lived swamp and marsh systems that endured through apparently ideal conditions for lush vegetation growth.

East of the Waterpocket Fold and in the Henry Mountains equivalent

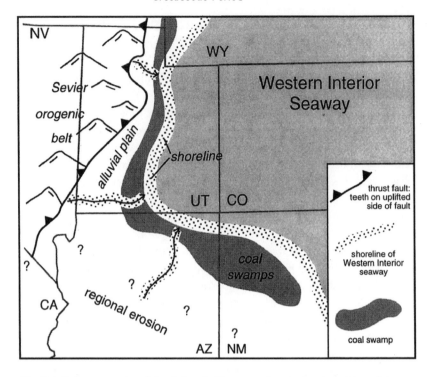

Fig. 5.6. Paleogeography of the Colorado Plateau region during deposition of the John Henry Member of the Straight Cliffs Formation on the Kaiparowits Plateau. Coal swamps dominated the depositional setting in the Kaiparowits area while the Blue Gate Shale was being deposited in deeper water of the seaway a short distance to the east, in the Henry Mountains area. After Roberts and Kirschbaum 1995.

rocks make up the Blue Gate Shale Member of the Mancos Shale (fig. 5.5). The Blue Gate consists of more than 1,000 feet of blue-gray, deep-marine shale. A sparse assemblage of fossils suggests deposition at water depths in the range of 200 to 400 feet.

During John Henry/Blue Gate deposition in southern Utah sluggish rivers meandered through vegetated floodplains to meet the sea within densely vegetated coastal swamps (fig. 5.6). These impenetrable bogs were bordered to the east by beach and shallow-marine environments where shelled organisms flourished. The sea deepened rapidly eastward, where black, organic-rich mud rained to the sea floor, eventually to become the Blue Gate Shale.

Revitalized rivers were next to sweep across the Kaiparowits region, blanketing the preceding coastal lowland with coarse sand and gravel. The approximately 200 feet of sandstone and conglomerate deposited by these east-flowing rivers make up the Drip Tank Member, the final event in the history of the Straight Cliffs Formation (fig. 5.5).

Equivalent coal beds and fossil-rich marine sandstone in the Henry Mountains and surrounding desert constitute the Emery Sandstone, a tongue of coarser sediment that extends eastward into the Mancos Shale (fig. 5.5). As you drive through the shale badlands on Highway 24, between the settlements of Hanksville and Caineville, the Emery forms the obvious brown sandstone cliffs that dominate the skyline.

Still farther east, deep-sea conditions endured, and the constant rain of mud onto the sea floor persisted as it had for the previous 10 million years. So the ordered pattern of depositional environments (high- and low-energy rivers, coastal plain, shoreline, shallow and deep marine) that lasted through the Cretaceous Period continued, but shifted eastward. Thus, the tropical swamplands at one time located in the Kaiparowits area (John Henry Member) were displaced about 75 km to the northeast into the Henry Mountains region (Emery Sandstone; fig. 5.5).

The exact impetus for the eastward shift of depositional systems is uncertain, but given the regional setting several reasonable scenarios can be entertained. One major factor is recurring upheaval in the Sevier belt to the west. A sharp pulse of renewed uplift could have launched a wave of sediment into the basin to spread across the coastal plain, shifting the entire mosaic of environments eastward. A simple drop in sea level could have had the same effect, however. In reality, the eastward step of all these systems probably was caused by some combination of these and other factors, including changes in climate and basin subsidence rates, to name but a few.

Wahweap Formation and Masuk Member of the Mancos Shale ≈

The Wahweap Formation records yet another modification to the Cretaceous landscape of southern Utah. While the sea was confined to the east about 1,500 feet of sandstone, mudstone, and siltstone were deposited in the Kaiparowits region by low-energy rivers. As these rivers wound lazily across the broad coastal plain, they were reduced to carrying only fine sand and mud. This deviation from the coarse-grained, high-energy rivers of the previous Drip Tank Member points to a significant energy drop in the Wahweap rivers. Coarser sediment was occasionally transported in these streams during exceptional flood events. During floods flow was turbulent enough to erode cohesive chunks of mud from the riverbanks and floodplain. After being transported a short distance these mud clasts eventually came to rest on the channel floor. Clay clasts of this origin, as well as broken fragments of dinosaur bone, are abundant in the large sandstone lenses that fill the ancient river channels. These Cretaceous channels are beautifully displayed along the cliffs of the Wahweap Formation, in the higher, interior region of the Kaiparowits Plateau.

The top 300 feet of the Wahweap in the Kaiparowits area are considerably coarser than lower parts. The vertical change from fine sandstone and mudstone to coarse sandstone and conglomerate marks an increase in the gradient of the rivers. Rejuvenation probably was linked to yet another phase of uplift to the west.

Equivalent strata to the east form the Masuk Member of the seemingly endless Mancos Shale succession (fig. 5.5). The Masuk is best exposed in the moundlike badlands that fringe the Henry Mountains to the northwest; here it consists of more than 600 feet of olive-brown shale and sandstone. Its exact origin confounded early geologists, who wavered between continental and marine interpretations. It was agreed that sedimentary structures hinted at a low-energy system, but this could fit either interpretation. Eventually, the recovery of fossils from the Masuk aided in the interpretation of a continental origin, although it still is not a clear-cut case. Fossil palynomorphs, those extremely useful microscopic pollen and spores from the abundant Cretaceous foliage, provide the best evidence for a nonmarine setting. Similarly, fossil gastropods, crocodilian teeth, fish scales, and turtle fragments suggest fresh water. Rare fossils, however, could be taken to indicate brackish water conditions and the mixing of fresh water and sea water. Recent interpretations, based on the fossil assemblage, rock types, and sedimentary structures, suggest deposition in very low-energy meandering rivers, probably quite near their mouths. The shoreline likely lay only a short distance to the east, enabling sea water to wash inland during storms or extreme high tides. The rapid eastward transition from the Masuk into a monotonous sequence of deep-marine shale likewise suggests such a setting.

Kaiparowits Formation ≈

The Kaiparowits Formation, named for its extensive exposures along the northwest side of the Kaiparowits Plateau, is a thick sequence (> 2,800 feet) of carbonaceous sandstone, shale, and siltstone that weathers into blue-gray slopes. These badlands define the northwest margin of the plateau and form a foreboding gray landscape. Westbound travelers on Highway 12, between the towns of Escalante and Henrieville, view this undulatory land-

Photo 5. Extensive badlands of the Cretaceous Kaiparowits Formation in "the Blues" on the west edge of the Kaiparowits Plateau. The formation consists of gray and brown sandstone, siltstone, and shale deposited by low-energy rivers that meandered eastward toward a nearby seaway. Powell Point in the background is composed of overlying Tertiary strata.

scape when driving down off the west side of the plateau near Henrieville. These extensive badlands are called "the Blues," for their steely blue-gray color (photo 5).

The Kaiparowits Formation cuts deeply into the underlying Wahweap Formation, as much as 45 feet in some places. The contact between the two units is marked by differences in their resistance to erosion. The upper part of the Wahweap is characterized by vertical brown cliffs, whereas the overlying Kaiparowits forms gray slopes, sparsely dotted with piñon and juniper trees.

The fine-grained sediments of the Kaiparowits Formation were strewn across the floodplain by relatively placid meandering rivers. The resulting rocks contain abundant fossil bones, including skeletons of turtles and crocodiles and parts of dinosaurs, fish, and mammals. Freshwater molluscs and petrified wood are also common. The preservation of a surprisingly wide variety and abundance of terrestrial fossils is a result of the low-energy depositional environment that dominated this period. Potential fossil material is more likely to be preserved in low-energy settings, largely because the dead organisms are not apt to be transported far from where they died. The transport of bones, shells, and wood in turbulent streams tends to destroy such material or at least render it unrecognizable, due to the abrasion and breakup of less-resistant organic material during constant collisions with sand, pebbles, and cobbles in the stream channel. Thus, when compared with other Cretaceous strata in the region, the Kaiparowits does not necessarily represent a time of exceptionally abundant life, but instead portrays ideal conditions for preservation after death.

The headwaters of the Kaiparowits drainage, like those of all the rivers of the previous 75 million years in the region, were in the Sevier belt to the west. These rivers flowed eastward, uninterrupted, to the Western Interior sea. Up to this time the vast foreland basin remained virtually unbroken by any topographic relief, except for the gentle eastward slope that extended from the abrupt edge of the Sevier mountain front to the shores of the sea.

Correlative rocks in the Henry Mountains region have been removed by erosion but were probably very similar to the Kaiparowits rocks. Similar-age rocks in the Book Cliffs, along the Utah-Colorado border, show a transition from sluggish stream-related deposits to marine-influenced sediments, indicating that the shoreline was situated in this area.

Canaan Peak Formation ≈

The Canaan Peak Formation is the conclusive Cretaceous unit in southern Utah and, in fact, probably extends across the Cretaceous-Tertiary time boundary. The formation has a limited exposure on the Kaiparowits Plateau, where it forms an erosional remnant that caps its namesake, Canaan Peak. It is also well exposed along the precipitous margins of the Table Cliffs Plateau, which essentially is a higher, northern continuation of the Kaiparowits Plateau.

The Canaan Peak Formation consists of up to 700 feet of conglomerate, sandstone, and mudstone. These rocks express an increase in energy and the development of a braided stream system capable of moving a coarse load. Sedimentary structures in sandstone and conglomerate clearly indicate deposition by east-northeast–flowing streams. This unwavering flow direction reflects little change from the regional geography of the previous tens of millions of years.

A close look (microscopic actually) at the composition of sand grains that make up the Canaan Peak, however, reveals some surprising changes. Sand in earlier Cretaceous sediments was dominated by quartz grains with lesser amounts of chert particles. These compositions point to a source of older sedimentary rocks that were thrust upward to form mountains to the west. The Canaan Peak Formation, while also derived from these Sevier highlands, contains 40 to 50% grains of volcanic origin. Geologist Patrick Goldstrand (1994), who documented this abrupt change, interpreted it to reflect erosion deep into the hinterland of the Sevier belt. In fact, chemical "fingerprinting" of these grains has shown that they are virtually identical to Jurassic volcanic rocks found today only in southeast California. This implies that by the time of Canaan Peak deposition large drainage systems had penetrated deeply into the mountain range, tapping new and unique rock types.

The Close of the Cretaceous Period ≈

The close of the Cretaceous Period worldwide was punctuated by an unparalleled global extinction event that obliterated two-thirds of the known species. This event marks the demise of the dinosaurs' 140-million-year reign on earth and closed the curtain on the Mesozoic Era.

Although still controversial, convincing evidence points to a meteor collision with the earth as the trigger for this extinction. Over the last ten years huge amounts of data have been collected from Cretaceous-Tertiary boundary rocks around the world. In an overwhelming number of instances these rocks contain compelling evidence for a major meteor impact.

The site of the meteor impact has been narrowed down to the Yucatán Peninsula of Mexico. The probable impact crater, named the Chicxulub Crater for a nearby village, is about 180 km across and lies partly on the peninsula and partly beneath the Caribbean Sea. The combination of erosion and sedimentation over the last 65 million years, however, has rendered the crater unrecognizable at the surface. The crater can be discerned only by drill-hole information and subsurface imaging techniques. The 65-million-year age, which coincides with the end of the Cretaceous Period, coupled with the immense size of the Chicxulub Crater, appears to establish this impact as the main factor in this unprecedented extinction event.

In a more local sense, the end of the Cretaceous on the Colorado Plateau also saw some dramatic changes. The Western Interior seaway receded from the region for the last time, not to be seen on the Plateau again. Tectonic activity in the Sevier orogenic belt ceased, although remnant highlands endured through much of the early Tertiary Period. Compressional stress instead was transferred eastward, into the foreland, where previously only the extensive sedimentary basin lay. This migration of compression created uplifts in the form of both subtle upwarps and full-fledged mountain ranges. In

southern Utah uplifts were in the form of mild upwarps that served to disrupt the previous west-to-east drainage system and produce a complex system of rivers and large lakes. By contrast, sharp uplifts developed in present-
day Colorado, New Mexico, and Wyoming, forming the dramatic
mountain ranges that are the modern Rocky Mountains. The eastward migration of tectonic activity and the deformation of the previously stable
foreland mark the beginning of what is known as the Laramide orogeny.

References ≈

Aubrey, W. M. 1996. Stratigraphic architecture and deformational history of Early
    Cretaceous foreland basin, eastern Utah and southwestern Colorado. In *Geology
    and resources of the Paradox Basin*, ed. A. C. Huffman, Jr., W. R. Lund, and L. H.
    Godwin, pp. 211–20. Guidebook 25. Salt Lake City: Utah Geological
    Association.

Bowers, W. E. 1972. The Canaan Peak, Pine Hollow, and Wasatch Formations in the
    Table Cliff region, Garfield County, Utah. *United States Geological Survey Bulletin*
    1331-B:B1–B39.

Cobban, W. A., and J. B. Reeside, Jr. 1952. Correlation of the Cretaceous formations
    of the western interior of the United States. *Geological Society of America Bulletin*
    63:1011–44.

Cole, R. D. 1987. Cretaceous rocks of the Dinosaur Triangle. In *Paleontology and ge-
    ology of the Dinosaur Triangle*, ed. W. R. Averitt, pp. 21–35. Grand Junction:
    Museum of Western Colorado.

Currie, B. S. 1997. Sequence stratigraphy of nonmarine Jurassic-Cretaceous rocks,
    central Cordilleran foreland-basin system. *Geological Society of America Bulletin*
    109:1206–22.

Eaton, J. G. 1991. Biostratigraphic framework for the Upper Cretaceous rocks of the
    Kaiparowits Plateau, southern Utah. In *Stratigraphy, depositional environments,
    and sedimentary tectonics of the western margin, Cretaceous Western Interior
    Seaway*, ed. J. D. Nations and J. G. Eaton, pp. 47–64. Special Paper 260. Boulder:
    Geological Society of America.

Eaton, J. G., P. M. Goldstrand, and J. Morrow. 1993. Composition and stratigraphic
    interpretation of Cretaceous strata of the Paunsaugunt Plateau, Utah. In *Aspects of
    Mesozoic geology and paleontology of the Colorado Plateau*, ed. M. Morales, pp.
    153–62. Bulletin 59. Flagstaff: Museum of Northern Arizona.

Elder, W. P., and J. I. Kirkland. 1993. Cretaceous paleogeography of the Colorado
    Plateau and adjacent areas. In *Aspects of Mesozoic geology and paleontology of the
    Colorado Plateau*, ed M. Morales, pp. 129–52. Bulletin 59. Flagstaff: Museum of
    Northern Arizona.

Fillmore, R. P. 1991. Tectonic influence on sedimentation in the southern Sevier fore-land, Iron Springs Formation (Upper Cretaceous), southwestern Utah. In *Stratigraphy, depositional environments, and sedimentary tectonics of the western margin, Cretaceous Western Interior Seaway*, ed. J. D. Nations and J. G. Eaton, pp. 9–26. Special Paper 260. Boulder: Geological Society of America.

Goldstrand, P. M. 1992. Evolution of Late Cretaceous and Early Tertiary basins of southwest Utah based on clastic petrology. *Journal of Sedimentary Petrology* 62:495–507.

———. 1994. Tectonic development of Upper Cretaceous to Eocene strata of south-western Utah. *Geological Society of America Bulletin* 106:145–54.

Lageson, D. R., and J. G. Schmitt. 1994. The Sevier orogenic belt of the western United States: Recent advances in understanding its structural and sedimentologic framework. In *Mesozoic systems of the Rocky Mountain region, USA*, ed. M. V. Caputo, J. A. Peterson, and K. J. Franczyk, pp. 27–64. Denver: Rocky Mountain Section, Society for Sedimentary Geology.

Peterson, F. 1969. Four new members of the Upper Cretaceous Straight Cliffs Formation in the southeastern Kaiparowits region, Kane County, Utah. *United States Geological Survey Bulletin* 1274-J:J1–J28.

Peterson, F., R. T. Ryder, and B. E. Law. 1980. Stratigraphy, sedimentology and re-gional relationships of the Cretaceous System in the Henry Mountains region, Utah. In *Henry Mountains Symposium*, ed. M. D. Picard, pp. 151–70. Publication 8. Salt Lake City: Utah Geological Association.

Roberts, L. N. R., and M. A. Kirschbaum. 1995. Paleogeography of the Late Cretaceous of the western interior of middle North America: Coal distribution and sediment accumulation. *United States Geological Survey Bulletin* 1561. Washington, D.C.: United States Geological Survey. 115 pp.

Schwans, P. 1988. Depositional response of Pigeon Creek Formation, Utah, to initial fold-thrust belt deformation in a differentially subsiding foreland basin. In *Interaction of the Rocky Mountain foreland and the Cordilleran thrust belt*, ed. C. J. Schmidt and W. J. Perry, Jr., pp. 531–56. Memoir 171. Boulder: Geological Society of America.

Shanley, K. W., and P. J. McCabe. 1991. Predicting facies architecture through se-quence stratigraphy—An example from the Kaiparowits Plateau, Utah. *Geology* 19:742–45.

6

Tertiary Period

The Tertiary Period, which ranges from 66 to 1.6 m.y.b.p., saw the initiation of both geological and biological events that have shaped the earth today. The extinction of the dinosaurs at the end of the Cretaceous cleared the way for the evolution of new and larger forms of mammals that continued to evolve through the Tertiary and ensuing Quaternary Period. As in the rest of the world, the Tertiary on the Colorado Plateau marks the inauguration of events that molded the modern landscape. During the latest part of the Cretaceous Period the Western Interior seaway pulled out of the region for the last time. Simultaneously, the belt of Sevier deformation that occupied western Utah and Nevada suddenly shifted east, disrupting the basin that had dominated western North America for the previous 90 million years. This mountain-building episode, which differed substantially in time, space, and style from the Sevier event, is called the *Laramide orogeny*. Laramide deformation, which reached as far east as the present-day Front Range in Colorado, was active from 65 to about 50 m.y.b.p.

As the disconnected Laramide mountain ranges were pushed up throughout the foreland, isolated basins developed between them, acting as giant traps for the detritus shed from their flanks. These lofty highlands were thrust upward in New Mexico and Colorado to form the southern Rocky Mountains.

While these steep mountain fronts were rising, southern Utah and the rest of the Colorado Plateau evolved somewhat differently. Although affected by Laramide deformation, the Colorado Plateau expressed it in a subtle but no less significant fashion. Rather than breaking up into jagged mountain ranges, the thick, stable crust of the plateau region was cut by lengthy, deep faults that caused the thick overlying sedimentary succession to drape over them. These deep-seated basement-involved faults produced the distinctive monoclinal folds that are scattered over the region today, including the Waterpocket Fold, San Rafael Swell, and Escalante monocline (fig. 6.1).

It was the early Tertiary monoclines that guided the location and style of Tertiary sedimentation in southern Utah. The upwarped part of these great

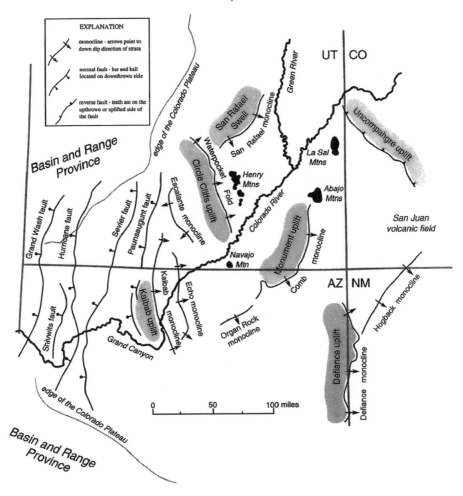

Fig. 6.1. Map showing the Tertiary structural geology of southern Utah and the surrounding region, with an emphasis on monoclines, normal faults along the west margin of the Colorado Plateau, and the igneous intrusive rocks (in black). After Peterson and Turner-Peterson 1989.

flexures acted as sediment sources, while adjacent low-lying areas collected this sediment. Tertiary sedimentary rocks are absent east of the Waterpocket Fold and to the west are first seen on the flanks of the Aquarius Plateau. Westward from there the colorful orange, pink, and white Tertiary strata become increasingly conspicuous as they cap the high promontories of the Table Cliffs and Mt. Powell and the Paunsaugunt Plateau, where the layers are carved into the extraordinary fins and pinnacles of Bryce Canyon National Park. Still farther west, across the Sevier River valley on the Markagunt Plateau, the same rocks form Cedar Breaks

National Monument (fig. 6.1). The pastel-colored escarpment that delineates the southern outline of the Paunsaugunt and Markagunt Plateaus constitutes the Pink Cliffs, the final step in the Grand Staircase.

The Tertiary also witnessed the introduction of igneous activity to the Colorado Plateau, something that had not been recorded in the previous 600 million years. The first flare-up of magma, which formed the Abajo, La Sal, and Henry Mountains, occurred over the period of about 32 to 23 m.y.b.p. These isolated alpine islands are the result of intrusive magma that cooled slowly beneath the surface. In all three instances, as the magma ascended to shallow levels in the crust it squeezed in between the sedimentary layers. This lateral injection forced the overlying strata upward, forming the mountain-sized blisters of resistant igneous rock that have been unearthed by erosion today.

Following close on the heels of the intrusive Henry Mountains were the fluid lava extrusions that form the dark palisades of the Aquarius Plateau. The 21.8 *Ma* (mega-annum or million years) age for the flow has been indirectly obtained from radiometric dating of similar, probably correlative rocks in the Tushar Mountains about 50 km to the west (Anderson and others 1975).

After a relatively quiet interval fluid lava once again poured from great fractures or fissures in the crust. The remnants of these younger flows form the high points of the Aquarius Plateau and its northern sister, Thousand Lakes Mountain. This basalt is part of a wave of magmatic activity that swept across the region from 6.4 to 4.0 Ma (Nelson and Tingey 1997).

Tectonic Setting: The Laramide Orogeny ≈

For years following the general acceptance of plate tectonics theory geologists puzzled over a cause for the Laramide orogeny. The initiation of the Laramide in the latest Cretaceous slightly overlapped in time with the demise of the Sevier orogeny. Yet Sevier deformation was confined to the region west of western Utah, while the onset of the Laramide was marked by a pronounced eastward shift of deformation throughout Utah and as far east as central Colorado. Moreover, Laramide uplift occurred along high-angle reverse faults that cut deep into the crust, thrusting ancient plutonic and metamorphic basement rocks to the surface, a style known as "thick-skinned" deformation. In contrast, basement rocks were rarely involved in the earlier Sevier highlands, which instead were characterized by thin-skinned low-angle thrust faults that were confined to sedimentary rocks of the upper crust.

While some have argued that the Laramide orogeny merely represents a continuation of Sevier deformation, several lines of evidence point to a real change in the plate tectonic configuration of western North America at the end of the Cretaceous. First, the chain of volcanoes to the west that had been erupting ash and lava for more than 100 million years was extinguished in the Late Cretaceous. Additionally, it is difficult to transmit compression of the magnitude required for the Laramide upheaval across the vast expanse of western North America. The simple subduction along the Pacific/North America plate boundary that is invoked for the Sevier orogeny probably would not produce *that* much compression. A plate tectonic configuration analogous to that for Cretaceous western North America is shown in figure 2.5. Clearly, an alternative mechanism or a variation on this earlier geometry is required to explain the changes seen in the Tertiary.

In 1978 William Dickinson and Walter Snyder proposed a plate tectonic model for the Laramide orogeny that handily explained both the shutdown of volcanism and the eastward migration of deformation. Their model had the additional strength of a modern analogue in a segment of the extensive Andes Mountain chain in western South America. In other words, the modern equivalent of the Laramide orogeny now is taking place in Argentina. While the model of Dickinson and Snyder was preliminary and thus subject to modification or even rejection as more was learned about the Laramide, it has stood the test remarkably well and remains the best explanation for the orogenic event.

In this model the oceanic crust that floored the Pacific Ocean was moving east and was being subducted or overridden by the continental plate of North America. Throughout the Cretaceous the oceanic plate descended at a steep angle, eventually melting as it reached a relatively deep level. As it began to melt the less dense molten rock crept upward, to erupt as volcanoes or slowly cool beneath the surface to crystallize into the granite exposed today in the Sierra Nevada range in eastern California. It was postulated that the subducting oceanic plate assumed a flatter angle toward the end of the Cretaceous, however, allowing it to push eastward beneath the North American plate and extend far inboard (fig. 6.2). In this setting the plate would not dive to the great depths it had earlier attained and thus would not melt. This explains the interruption in volcanic activity. As the oceanic plate scraped its way eastward beneath North America, tremendous shear stresses were generated between the two slabs. Shear stress along the bottom of the thick continental crust was transmitted upward in the form of compression, thrusting great wedges of basement rock skyward to form the spectacular Laramide highlands seen today.

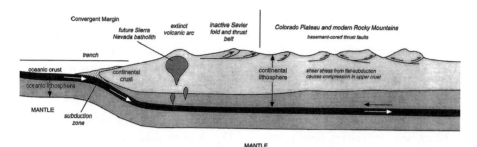

Fig. 6.2. General cross section through the crust and upper mantle of western North America during the Tertiary showing the dramatic change in the geometry of the subducting oceanic crust. An increase in the rate of production of oceanic crust at the mid-ocean ridge (not shown) increased the temperature of the subducting crust. This in turn caused the subducting crust to decrease its angle as it merged with the North American continent. This "flat-slab" subduction terminated volcanism in the Sierra Nevada volcanic arc because the oceanic crust never reached the depth required for it to begin melting. More importantly, shear stress at the base of the continental lithosphere was greatly increased far into the continent from the subduction zone as the oceanic crust scraped along its base. These shear stresses caused faulting in basement rocks in the upper crust, breaking up the earlier Cretaceous foreland basin to produce the monoclines of the Colorado Plateau and the modern southern Rocky Mountains of New Mexico, Colorado, and Wyoming.

Southern Rocky Mountains ≈

There are profound differences in the style of Laramide uplift observed between the southern Rockies and the Colorado Plateau, despite their being shaped by the same stresses. These discrepancies can be blamed on regional variations in the strength and stability of the crust. Many of the sharp, mountainous uplifts in the southern Rockies were caused by the reactivation of preexisting crustal weaknesses such as faults. Whenever stresses, large or small, act on a heterogeneous body of rock, such flaws will be exploited. In one of the best-documented examples, the modern Rocky Mountains in Colorado coincide with the Pennsylvanian-age uplifts that dominated the landscape 300 million years earlier. Subsurface studies on the boundaries of the modern Front Range and the Uncompahgre Plateau show that much of the Laramide deformation was along faults that also were active in the Pennsylvanian (fig. 6.1). In addition, some geologists have suggested that these faults have a still earlier origin, possibly as far back as 700 million years ago.

Colorado Plateau ≈

Since geologists first encountered the bare rock of the Colorado Plateau, monoclines have been reported as the most conspicuous and abundant structural features in the region. Although John Wesley Powell described these curious upwarps in 1873, the term "monocline" was introduced by Charles Dutton in his 1882 report, *The Physical Geology of the Grand Canyon District*. By definition, a monocline is a large-scale, steplike fold whose single limb dips in one direction. This ramp of rock is bounded on both sides by relatively flat-lying strata (fig. 6.3).

In the past the origin of monoclines has provoked much speculation, partly because these monstrous ramps are so plentiful over the Plateau region and scarce elsewhere. Many of these upwarps snake more than 100 miles across the open landscape. The Waterpocket Fold, which constitutes a large part of Capitol Reef, extends for 70 miles, while the neighboring San Rafael Swell is 60 miles long. In addition, more than 7,000 feet of uplift has taken place on the Waterpocket Fold, a magnitude comparable to the mountainous uplifts of the Rocky Mountains to the east. Today it is generally agreed that these great monoclines are underlain at depth by reverse faults in the basement rock and the thick overlying strata have, for the most part, "gone along for the ride" (fig. 6.3). The compression required for the generation of these regional faults originated with the Laramide orogeny.

The greater stability of the Colorado Plateau relative to the rest of western North America may be revealed in a whirlwind tour of its geologic history. Through most of the Paleozoic Era the Plateau remained a broad shelf covered by a shallow sea. The region never felt the episodic, rapid downwarping that affected the sea floor to the west. The Plateau also remained immune to the highlands of crumpled rock and volcanoes that intermittently haunted its western margin. Toward the end of the Paleozoic, as the Ancestral Rockies rose to the east, the Plateau remained undisturbed except for localized subsidence and the development of some broad upwarps. Through most of the Mesozoic, as orogenic activity escalated, the region experienced only continual subsidence, receiving the sediments that form the magnificent strata exposed today. Even the Sevier orogeny, probably the most vigorous mountain-building event up to that point, halted at the western margin of this stable region.

As in the southern Rockies, monoclines on the Colorado Plateau were controlled by preexisting weaknesses in the crust. Abundant evidence suggests that modern monoclines also experienced uplift during the Permian, Triassic, and Jurassic Periods. The Circle Cliffs, which mark the crest of the Waterpocket Fold, experienced mild uplift during the Early Triassic and again in Middle to Late Jurassic time (Huntoon and others 1994; Peterson 1994). Today the Monument Upwarp in southeast Utah forms a broad platform that is bounded to the east by the Comb Ridge monocline (fig. 6.1). Jurassic strata thin or pinch out against this highland, pointing to a

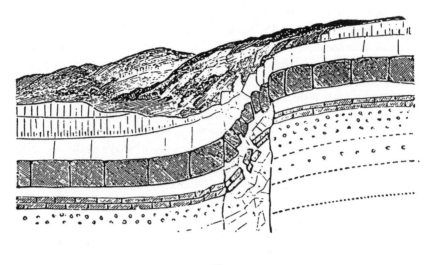

a

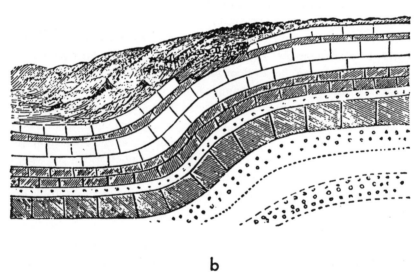

b

Fig. 6.3. Two possible origins for the monoclinal folds of the Colorado Plateau region as proposed by John Wesley Powell. *A* is a normal fault in which broken rock
fills the fault zone, giving the illusion of continuous strata. *B* shows a simple monoclinal fold. While both interpretations are possible, most of the monoclines on the
Colorado Plateau have more recently been shown to be underlain by reverse faults in
the basement. From Powell 1873.

similar configuration about 175 million years ago (Blakey 1994; Peterson 1994). The Kaibab monocline apparently was also active in the Early Triassic and Late Jurassic. This monocline is of particular interest because it is one of the few whose faulted roots have been excavated by erosion, in this case in the bottom of the Grand Canyon. Detailed studies have shown that this fault was first active as a normal fault during extension of the crust in Precambrian time, more than 570 million years ago. Subsequent reactivation of the fault during Early Tertiary compression, however, produced an opposite or reverse movement (Huntoon 1993). The Kaibab monocline, one of the lengthiest on the Colorado Plateau at 130 miles, extends northward into southern Utah to form the Cockscomb. Here the sharp fold marks the southwest boundary of the Kaiparowits Plateau in the Grand Staircase–Escalante National Monument.

Tertiary Stratigraphy ≈

As we have seen time and again, the sedimentary record holds the most complete account of tectonic disturbance for a region. The Tertiary record of southern Utah is no different. The absence of Tertiary strata east of the Waterpocket Fold, however, forces our attention toward the Aquarius Plateau and points west, to the area labeled the High Plateaus region by Dutton in 1880. Much of this landscape is held up by a veneer of resistant volcanic caprock, a significant factor in the protection of soft sedimentary layers that reside beneath. These Tertiary sedimentary rocks are widely exposed along the rims and dissected interiors of the Table Cliff, Paunsaugunt, and Markagunt Plateaus (fig. 6.4). Remnants of these rocks also cling to isolated summits of the Kaiparowits Plateau to the south and include its high point, Canaan Peak. These scattered outliers provide valuable clues to the distribution of Tertiary strata before widespread erosion.

The sedimentary history for the Colorado Plateau resumes with the strata that overlie the Canaan Peak Formation, which straddles the Cretaceous-Tertiary boundary (see chapter 5). Tertiary sediments of the High Plateaus range from Paleocene to Oligocene age. Time divisions commonly used for the Tertiary differ from the standard periods used to subdivide the older Paleozoic and Mesozoic Eras. Periods may range up to 65 million years, whereas the epochs that subdivide the Tertiary rarely exceed 20 million years. As we edge closer to the present, the more complete preservation of the strata with their associated fossil record permits a finer-scale of division into the shorter epochs.

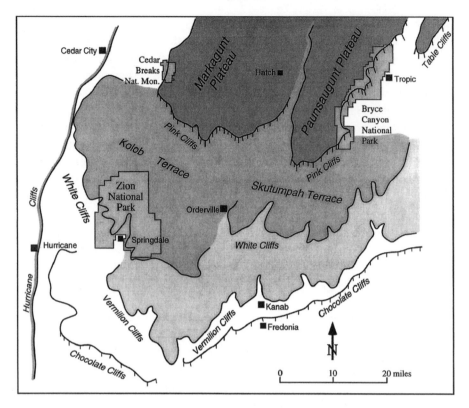

Fig. 6.4. Geography of the High Plateaus region in southwest Utah. This region defines the west margin of the Colorado Plateau. West of the Hurricane Cliffs lies the Basin and Range physiographic and geologic province.

Grand Castle Formation ≈

The Paleocene Grand Castle Formation only recently was formally defined by geologists Patrick Goldstrand and Doug Mullett (1995), although the rocks that compose it have been known since they were first described by Dutton in 1880. The coarse conglomerate and sandstone of the Grand Castle had previously been placed with several different formations, depending on the location. On the Table Cliff Plateau these sandstones and conglomerates were considered part of the underlying Canaan Peak Formation, which contains similar conglomerate. The Grand Castle Formation, as defined by Goldstrand and Mullett, is differentiated from underlying strata by a change in the makeup of the pebbles and cobbles in conglomerate beds; volcanic clasts in the Canaan Peak are abruptly replaced by limestone pebbles in the overlying Grand Castle conglomerates.

Along the western Markagunt Plateau Grand Castle boulder conglomerates are easily differentiated from fine sandstone of the underlying Cretaceous Iron Springs Formation. Here the Grand Castle originally was grouped with overlying siltstone and limestone to form the Claron Formation. The stratigraphic revision now places the conglomerate and sandstone in the Grand Castle Formation, leaving the finer-grained strata as the sole constituent of the overlying Claron.

The Grand Castle thins rapidly to the south and is absent on the Paunsaugunt and southern Markagunt Plateaus. This suggests the presence of a mild positive area during deposition, probably an early disclosure of Laramide compression (Goldstrand and Mullett 1995). This feature was not high enough to act as a significant sediment source, but it apparently did deflect the course of the Grand Castle drainage, halting any southward expansion.

The Grand Castle Formation is especially noteworthy in the tectonic history of the western Colorado Plateau because it traces the conclusion of the Sevier orogeny and the rise of Laramide disruption across the foreland. The formation was deposited by east-flowing braided streams driven by the last gasp of uplift in the Sevier belt. To the west Grand Castle conglomerate overlaps what is interpreted as the final thrust fault to develop in the frontal Sevier highlands, but is unaffected by it (Goldstrand and Mullett 1995). Clearly, the Grand Castle postdates fault movement.

In contrast, the formation in the Table Cliff area to the east has been folded by Laramide activity, indicating that deposition preceded this pulse of deformation. Collectively these relations show that the Grand Castle was deposited after the Sevier orogeny, but before Laramide folding. The formation occupies a unique position and offers a geologic snapshot of that brief instant between these major mountain-building events of western North America.

Pine Hollow Formation ≈

Exposures of the Pine Hollow Formation are confined to parts of the Table Cliff Plateau, where it consists of 200 to 300 feet of mostly varicolored mudstone, siltstone, and limestone with scattered conglomerate at the base. The formation overlies Cretaceous strata or the Grand Castle Formation; all underlying strata were tilted then beveled flat before being mantled by the Pine Hollow sediments. It is overlain by the upper Claron Formation, which contains similar fine-grained rocks. The age of the Pine Hollow ranges from Early Paleocene to perhaps Middle Eocene, and it may be the time-equivalent of the lower Claron farther to the west. While the Claron and Pine Hollow Formations contain similar rock types and are approximately coeval, they were deposited in separate basins that were iso-lated by Laramide folds that began to rise throughout the foreland, dis-rupting the regional east-flowing drainage pattern that had endured for so long.

The small Pine Hollow basin lay nestled in a narrow downwarp barri-caded by Laramide uplifts. To the west was the Johns Valley anticline, a north-trending arch that formed along the west side of the present-day Table Cliff Plateau. To the east lay the phoenixlike Circle Cliffs, rising once again to the call of tectonism. The Pine Hollow served as a receptacle for detritus washed from both sides of the downfolded trough.

Pebbles in the basal conglomerate of the Pine Hollow Formation spilled into the basin from the west and the northeast (fig. 6.5). The pebble com-position is identical to that of the underlying Canaan Peak Formation. This discovery led Patrick Goldstrand (1992) to conclude that they had been re-cycled during erosion of the Canaan Peak off the Johns Valley anticline. Rivers that emanated from the anticline and the Circle Cliffs uplift were confined to the closed basin to feed the lake that occupied its center. Siltstone, mudstone, and limestone that dominate the formation accumu-lated in this shallow lake. The interpretation that the Pine Hollow Formation was deposited in a small closed basin and derived from local sed-iment sources provides the first clear evidence for partitioning of the once extensive foreland by Laramide uplifts. These events molded the pattern of basins and highlands that, although scarred by millions of years of erosion, endures today.

Claron Formation ≈

The Claron Formation is the most prominent and widespread of the Tertiary sedimentary rocks in southern Utah. Its chalky, multicolored strata make up the convoluted landscape of Bryce Canyon on the Paunsaugunt Plateau and Cedar Breaks National Monument on the adjacent Markagunt Plateau (fig. 6.4). In addition, the Pink Cliffs segment of the Grand Staircase takes its name from the pastels of the Claron. At Bryce the pink, red, orange, and white limestone and mudstone have been etched into a fantastic assemblage of hoodoos and spires, thanks to an eons-old collabo-ration between jointing and erosion (photo 6). These intricate forms radi-ate an astounding luminescence in the early morning or evening light and at times appear to glow from some kind of inner light source. Some of this can be attributed to the alpine setting of the Claron at an elevation of 8,000 to 9,000 feet. At these heights the cool, dry air transmits light much more efficiently, lending an increased clarity.

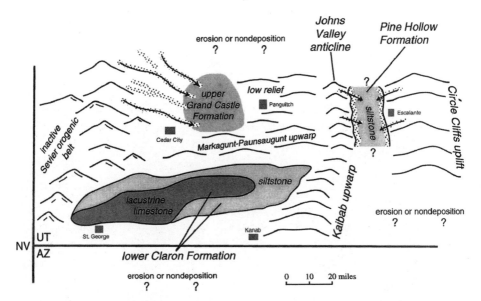

Fig. 6.5. Paleogeography of southwest Utah during the Late Paleocene and Early Eocene, showing the source highlands and depositional systems responsible for the fluvial upper Grand Castle Formation, lacustrine lower Claron Formation, and fluvio-lacustrine Pine Hollow Formation. After Goldstrand 1994.

The Claron is dominated by limestone with lesser amounts of mudstone cut by rare conglomerate lenses. It reaches a maximum thickness of about 1,500 feet at Cedar Breaks, one of the few localities where the complete thickness is preserved. The age of the Claron ranges from Late Paleocene to Eocene. As stated above, the basal Claron to the west is equivalent to the Pine Hollow Formation in the Table Cliff region, which collected under similar conditions but in a separate basin. The Pine Hollow, however, is overlain by the upper part of the Claron.

The Claron was deposited in a broad basin under relatively quiet conditions. Limestone accumulated in a large, shallow lake that covered much of southwest Utah. At its largest the basin extended east to the Circle Cliffs uplift, which was of sufficient relief to block any further expansion. The west boundary was defined by the front of the Sevier highlands, which, although dormant for 10 million years, maintained a topographic presence. In the northern part of today's Paunsaugunt and Markagunt Plateaus a broad floodplain bordered the lake. Primary sedimentary structures in these sediments have been obliterated by prolonged soil-forming processes so that many of these strata have been altered to paleosols. These floodplain soils are cut by lenses of conglomerate washed in from the north by the

Photo 6. The bizarre hoodoos of Bryce Canyon National Park, carved from the soft siltstone and more resistant limestone. These sediments were deposited in an extensive lake system that occupied the region during the Early Tertiary.

small streams that fed the lake. Mudflats developed along the lake margin near the Sevier and Circle Cliff highlands, suggesting fairly subdued topography in these areas.

Initially the Claron lake was replenished with water from the north and northwest; these streams originated in the declining Sevier highlands (fig. 6.5). The coeval Pine Hollow basin, immediately to the northeast, was partitioned from the huge Claron basin by the Johns Valley anticline, a subtle but effective barrier.

By Middle Eocene time (about 50 Ma) activity on the Laramide folds had shut down, and the basins began to fill slowly with fine sediment. Eventually the Johns Valley anticline was overtopped by the Claron lake, which flooded into the smaller basin, assimilating it into a single expanded basin. As sedimentation continued, the smaller intrabasinal highlands like the Johns Valley and Markagunt-Paunsaugunt upwarps were blanketed by limy mud. By the end of Claron deposition all the Laramide structures that had guided earlier depositional patterns were buried with sediment. Only the residual Sevier highlands and the more pronounced Circle Cliffs uplift remained to restrict the basin. The location and nature of the southern basin margin remain uncertain due to the absence of Tertiary sediments south of the High Plateaus.

The Claron is peculiar in its preservation of unique fossil insect burrows that have recently been identified in its paleosols (Bown and others 1997). Most fascinating in this assemblage is a fossilized ant nest exposed in cross section on a small cliff face just west of the High Plateaus region near the town of Parowan. The multiple chambers and horizontal passages in this exceptionally well-preserved nest are connected by vertical shafts, closely resembling the glassed-in "ant farms" cultivated in my youth and in classrooms across the United States today.

Another distinctive trace fossil consists of small cocoonlike cells that easily weather from the paleosols. These are interpreted by Thomas Bown and his colleagues as the traces of ground-dwelling bees and wasps. A different trace fossil, found exclusively in southwest Utah, is another small, burrowlike feature that has been named *Eatonichnus* (Bown and others 1997). The fossil is unlike anything observed in modern environments or in the rock record, but it is tentatively interpreted as the nest of a dung beetle. *Eatonichnus* is named for paleontologist Jeff Eaton, who has spent much of his career investigating the geology and paleontology of the southern Colorado Plateau, including the study of Cretaceous dinosaurs and mammals and Tertiary mammals.

Trace fossils, especially those found in ancient soils, may be useful paleoenvironmental indicators. Modern digging bees and wasps adapt best in semiarid to subhumid conditions, in well-drained soils. Collectively the insect traces found in the Claron point to a moderate climate with seasonal rainfall that produced alternating wet and dry seasons.

By the finish of Claron deposition the stage had been set for the evolution of the modern landscape. The prominent monoclines of the region, the Waterpocket Fold, San Rafael Swell, and the Kaibab and Escalante flexures, had all reached their zenith and were being dissected by the relentless hand of erosion—a process that continues today. In fact, as we have gleaned from the stratigraphic record, these great folds were being torn down even as they began to rise.

Some final ingredients of the present landscape were still absent at this time, including the intrusive and extrusive products of magmatism—the laccoliths of the Henry Mountains and the extensive plateau-forming lava flows to the west. It was not, however, a continuous procession from Laramide folding to igneous flare-up, as a relatively stable, quiet period separated the two events.

Tertiary Igneous Activity ≈

Henry Mountains Laccoliths ≈

The first major geologic event to affect southern Utah after the Laramide was the regional magmatic episode that formed the laccoliths of the Henry Mountains. Put simply, *laccoliths* are large blisters of magma that are intruded into originally flat-lying sedimentary rocks (fig. 6.6). Upon rising near the surface, the body of viscous magma squeezes laterally between two of the rock layers to form the characteristic mushroom shape. The base of the laccolith is flat, conforming to the top of the underlying sedimentary layer. Because this occurs at fairly shallow levels of the crust, pressure from the overlying layers is small, and the force of the intruding magma domes the overlying layers up to form the top of the laccolith. In cross-sectional view the mushroom shape is completed with a stemlike conduit that feeds the magma upward into the main body. With time, the overlying nonresistant sedimentary layers are removed by the combined erosive forces of water and gravity, leaving a resistant plutonic core that is flanked on all sides by the tilted remnants of the sedimentary rock.

The laccolith as a unique type of igneous intrusion was first recognized by Grove Karl Gilbert, one of the greatest geologists in American history. Under the direction of John Wesley Powell, Gilbert explored and examined the rocks of the Henry Mountains in 1875 and 1876. These field investigations quickly led to the publication of Gilbert's classic monograph *Report on the Geology of the Henry Mountains* in 1877. Laccoliths were first described and defined in this report. This notable contribution to the science of geology has insured that the Henrys, no matter how isolated, will endure as the classic example of laccoliths, known to geologists worldwide.

The Henry Mountains are also unique as the last mountain range in the United States to be named by American explorers and accurately placed on a map. This was due in great part to its rugged isolation. Even accounts from Powell's second Colorado River voyage in 1871 refer to the range as the "Unknown Mountains." The Henry Mountains were named by Powell for a friend and supporter of his early scientific explorations, Joseph Henry, who at the time was the director of the Smithsonian Institution.

Intrusion of the Henry Mountain laccoliths took place over the period of 31.2 to 23.3 Ma (Nelson and others 1992). Their age is based on radiometric dates calculated from the painstaking measurements of isotopes of the element argon ($^{40}Ar/^{39}Ar$). Individual mineral grains separated from the rock were analyzed by methods that are among the most accurate available for dating rocks today.

The age of the Henrys fits handily with those obtained from its sister

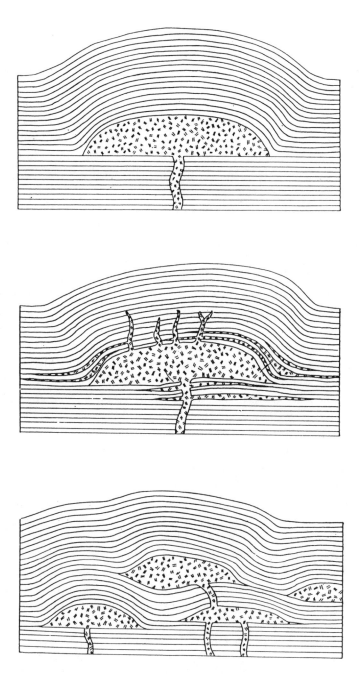

Fig. 6.6. Diagrams from G. K. Gilbert's classic study on the geology of the Henry Mountains showing interpretive cross sections of laccoliths and their various geometries and relationships with the sedimentary rocks that they intrude. From Gilbert 1877.

laccolithic mountains on the Colorado Plateau. The La Sal Mountains, which form the spectacular backdrop for the city of Moab, have yielded ages from 25.1 to 27.9 Ma, while the Abajo Mountains near the town of Monticello in southeast Utah range from 28.6 to 22.6 Ma (Nelson and others 1992). Only the discrete laccolith of Navajo Mountain, south of the Henrys, remains to be dated. It probably falls within the time frame of its expansive neighbors to the north, although a final verdict awaits further investigations. Similarities in composition, intrusive style and geometry, and age among these alpine islands of the Plateau point to a regional magmatic event and a single generating mechanism.

When trying to decipher the cause for this outbreak of magmatism, we must consider a link between the scattered clusters of laccoliths and volcanic fields to the east and west (fig. 6.1). The ages of these great igneous centers overlap slightly, suggesting a relationship. The San Juan volcanic field, which was active from about 30 to 20 Ma, forms the southeast margin of the Plateau. The towering San Juan Mountains of southwest Colorado, which make up this field, are pocked by eight or nine gigantic craterlike remnants or *calderas* that formed after the volcanoes had spewed huge volumes of incandescent ash and collapsed inward on themselves. Along the west margin of the Plateau lay the comparable Marysvale volcanic field, which vented lava and hot ash over a large part of west-central Utah. The Marysvale field was active over the period of about 26 to 21 Ma. When these volcanic ages are plotted on a map with the 31 to 23 Ma ages of the Henry, Abajo, and La Sal laccoliths an east to west sweep of igneous activity becomes apparent. A puzzling geologic pattern can once more be simply and effectively explained by plate tectonic relations.

The most likely cause for the abrupt outbreak of igneous activity on and around the Colorado Plateau is again the subduction process along the west margin of the continent, only this time with a twist. Sometime prior to this flare-up the mid-ocean ridge, which generates oceanic crust, was itself subducted, bringing the process to a grinding halt. When subduction ceased, the dense slab of oceanic crust that had been shoved beneath western North America slowly foundered and sank into the mantle, eventually to be recycled. As the oceanic slab broke up and collapsed, hot mantle material rose in its place, heating the overlying continental crust. The increased temperature caused parts of this crust to melt, producing magma. From about 30 to 20 Ma this magma made its way to or near the surface to form the laccoliths and volcanic fields that contribute to the landscape today.

The confinement of large-scale volcanic activity to the margins of the

Colorado Plateau, while the Plateau experienced only relatively small-volume laccolithic intrusions, bolsters the interpretation that the region is floored by thick stable crust. Thus only a relatively small quantity of magma was able to force its way upward to penetrate the dense, cool underside of the Plateau. Large-scale activity was pushed to the margins, where the crust had been weakened by hundreds of millions of years of deformation.

Following the intrusion of the laccoliths, but prior to the next volcanic event at about 6.0 Ma, the Plateau experienced only regional erosion. Precursors of today's drainages began to develop, cutting headward into the various Laramide upwarps and the recently uplifted Henry Mountains laccoliths; valleys were carved into low-lying areas and regions occupied by the soft Cretaceous shales. The throughgoing drainage of the modern Colorado River, however, was not established until later, and there is some debate over where these early rivers ultimately flowed. Our picture of this time is clouded by the apparent lack of sedimentary deposits of this age in the region. Because erosion dominated over deposition, the record is woefully incomplete, and we are forced to reconstruct the landscape from sparse evidence.

Basaltic Volcanism ≈

From about 6.5 to 3.4 Ma parts of southern Utah witnessed vast outpourings of fluid lava. This lava issued from numerous fissures to spread in thin flows across the land, producing the thin veneer of resistant basalt that shelters the Aquarius and Thousand Lakes Plateaus from erosion today. While erosion since this time probably has reduced the area covered by this basalt, it has also excavated and laid bare many of the fissures through which the lava escaped to the surface. These fissures are exposed as dikes intruded into sedimentary rocks of the San Rafael Group in the isolated desert between the Waterpocket Fold and the San Rafael Swell. *Dikes* are tabular or sheet-like bodies of intrusive igneous rock that are oriented vertically or at some high angle to the sedimentary bedding of the host rocks. The features considered here are more appropriately called *feeder dikes* since they "fed" the magma to the surface, acting as a conduit for its upward passage. Because the fracture-filling basalt is harder than the easily eroded San Rafael Group rocks that host it, the dikes are presented as resistant spines of black igneous rock against a backdrop of red sedimentary rock.

Although they are separated by up to 12 miles, a genetic relationship between the dikes and the plateau-capping lava flows is reinforced by similarities in age and composition. Basalt on the Aquarius Plateau remains to be dated, but flows on the neighboring Thousand Lakes Plateau yield a K-Ar date of

6.4 ± 0.4 Ma. Dikes to the east that have been dated provide ages ranging from 6.6 to 3.4 Ma. All the igneous rocks that fall within this age range, whether dikes or lava flows, are of basaltic composition with similar chemical signatures, although there are minor variations. This suggests a common source for the magma, deep in the mantle of the earth.

Geologists Stephen Nelson and David Tingey (1997) of Brigham Young University have studied the regional distribution and age of these young basalts and recognized patterns that provide significant clues to their origin. The flows and other volcanic features in southern Utah are part of an east-northeast–trending belt that stretches 400 km, from eastern Nevada to the San Rafael Swell area (fig. 6.7). An important aspect of this belt is the age of the basalts, which systematically decreases to the east. Basalt in Nevada, in the western part of the belt, began to erupt at about 15 Ma. The volcanism then slowly swept eastward at a rate of about 3.5 cm/year, until ending in the San Rafael Swell at about 3.0 Ma.

Subduction had ended tens of millions of years before, so different processes clearly have to be called upon to explain this new wave of activity. The linear belt of basalt and the eastward younging suggest a hot spot in the mantle, deep beneath the continental crust of North America. *Hot spots* are stationary points in the mantle that generate abnormally high heat flow—a sort of a blowtorch deep within the earth, heating the base of the crust. Hot spots are located below the constantly shifting tectonic plates that make up the uppermost part of the earth. As the hot spot generates basaltic magma by melting parts of the mantle, the magma begins its slow ascent through the crust. Linear age progressions are typical in hot spot–generated volcanism. Eastward younging of basalt in southern Utah and Nevada is caused by the west-southwest movement of the North American plate relative to the fixed hot spot. Current movement of the North American plate is documented to be to the west-southwest at a rate of 2.5 cm/year, a vector that fits well with the rate of volcanic migration for southern Utah. Similar patterns have been recognized for the Yellowstone region to the north and the San Francisco volcanic field to the south, in northern Arizona. The ages in both of these areas show that relatively recent volcanism migrated east-northeast through time.

Uplift of the Colorado Plateau ≈

Since the end of the Cretaceous Period 66 million years ago the Colorado Plateau has been uplifted about 12,000 feet. This number comes from the present elevation of Late Cretaceous shoreline deposits that occupy the

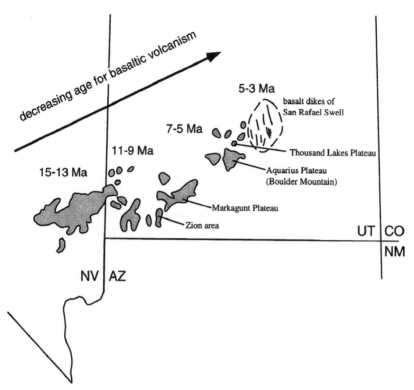

Fig. 6.7. Map showing the distribution of young basalts in southern Nevada and southern Utah and their approximate ages. Note the systematic decrease in age from west to east, suggesting the presence of a fixed hot spot in the mantle as a cause. Basalts in the Zion area are < 1.0 Ma and do not fit the 11–9 Ma grouping for that area. These younger basalts are probably related to the Hurricane fault and extension associated with the Basin and Range to the west. After Nelson and Tingey 1997.

loftier parts of the province. Several hypotheses have been proposed to explain the timing and cause of this uplift, but none so far is completely satisfactory. As a result, all are controversial, and there is no consensus among geologists working on the problem. The uplift of the Colorado Plateau and the attendant incision of deep canyons remain one of the great unsolved problems of North American geology.

Some geologists argue that plateau uplift was triggered by the Early Tertiary Laramide orogeny. It is this hypothesis that best explains *how* the uplift occurred—by large-scale tectonic activity. Opposition to this interpretation is based on the preservation of the colorful Mesozoic rocks that now blanket the Plateau. If these rocks had been shoved upward more than 50 million years ago, the thick sheets of sandstone would surely have been stripped from the region by prolonged erosion. Instead, Mesozoic strata

are preserved on the Colorado Plateau and have been eroded from surrounding areas.

Alternatively, many geologists argue that wholesale plateau uplift did not
commence until the Late Tertiary, about 5–6 million years ago. While the
exact impetus for uplift at this time remains obscure, abundant evidence
supports the timing. Much of this defense lies in the chronology of canyon-
cutting on the Plateau, specifically the genesis of the Grand Canyon.
Volcanic rocks as young as 6.0 Ma near Lake Mead, at the mouth of the
Grand Canyon, originally extended continuously across the present site of
the great gorge, indicating that the canyon had not initiated at this time.
Instead, it appears that closed basins with internal drainage dominated the
scene as Basin and Range extensional faulting chopped the region into
small, isolated basins. Within the Grand Canyon, however, are basalt flows
dated at 3.8 Ma perched 300 feet above the present level of the Colorado
River. These relations indicate the canyon had achieved its present status as
the regional trunk stream and had cut close to its present grade. The carving of the Grand Canyon from < 6.0 to about 3.8 Ma implies an incredible
rate of incision, even for the mighty Colorado River! Proponents of Late
Tertiary uplift attribute this incredible downcutting to uplift of the
Colorado Plateau, thereby accelerating the erosion rate over the vast upland. The biggest problem with this scenario is the lack of a clear mechanism for such large-scale uplift. No orogenic activity that is required to trigger such a large magnitude of uplift has been identified for Late
Miocene–Early Pliocene time. It is possible that the entire region of western North America was elevated during the Early Tertiary and that subsequent Late Tertiary extension dropped the Basin and Range, leaving the
Plateau as a relative highland. Work on this problem continues today.

References ≈

Anderson, J. J., P. D. Rowley, R. J. Fleck, and A. E. M. Nairn. 1975. *Cenozoic geology
of southwestern High Plateaus of Utah.* Special Paper 160. Boulder: Geological
Society of America. 88 pp.

Best, M. G., E. H. McKee, and P. E. Damon. 1980. Space-time-composition patterns
of late Cenozoic mafic volcanism, southwestern Utah and adjoining areas.
American Journal of Science 280:1035–50.

Blakey, R. C. 1994. Paleogeographic and tectonic controls on some Lower and Middle
Jurassic erg deposits, Colorado Plateau. In *Mesozoic systems of the Rocky Mountain
region, USA,* ed. M. V. Caputo, J. A. Peterson, and K. J. Franczyk, pp. 273–98.
Denver: Rocky Mountain Section, Society for Sedimentary Geology.

Bown, T. M., S. T. Hasiotis, J. F. Genise, F. Maldonado, and E. M. Brouwers. 1997. Trace fossils of Hymenoptera and other insects, and paleoenvironments of the Claron Formation (Paleocene and Eocene), southwestern Utah. *United States Geological Survey Bulletin* 2153-C:41–58.

Delaney, P. T., and A. E. Gartner. 1997. Physical processes of shallow mafic dike emplacement near the San Rafael Swell, Utah. *Geological Society of America Bulletin* 109:1177–92.

Dickinson, W. R., and W. S. Snyder. 1978. Plate tectonics of the Laramide orogeny. In *Laramide folding associated with basement block faulting in the western United States,* ed. V. Matthews III, pp. 355–66. Memoir 151. Boulder: Geological Society of America.

Dutton, C. E. 1880. *Geology of the High Plateaus of Utah.* Washington, D.C.: Dept. of Interior, U.S. Geographical and Geological Survey of the Rocky Mountain Region. 307 pp.

———. 1882. The physical geology of the Grand Canyon district. In *United States Geological Survey, 2nd annual report,* pp. 47–166. Washington, D.C.: Dept. of Interior.

Gilbert, G. K. 1877. *Report on the geology of the Henry Mountains.* Washington, D.C.: Dept. of Interior, U.S. Geographical and Geological Survey of the Rocky Mountain Region. 160 pp.

Goldstrand, P. M. 1992. Evolution of Late Cretaceous and Early Tertiary basins of southwest Utah based on clastic petrology. *Journal of Sedimentary Petrology* 62:495–507.

———. 1994. Tectonic development of Upper Cretaceous to Eocene strata of southwestern Utah. *Geological Society of America Bulletin* 106:145–54.

Goldstrand, P. M., and D. J. Mullett. 1995. The Paleocene Grand Castle Formation— A new formation on the Markagunt Plateau of southwestern Utah. *United States Geological Survey Bulletin* 2153-D:59–77.

Hunt, C. B., P. Averitt, and R. L. Miller. 1953. *Geology and geography of the Henry Mountains region, Utah.* Professional Paper 228. Washington, D.C.: United States Geological Survey. 234 pp.

Huntoon, J. E., R. F. Dubiel, and J. D. Stanesco. 1994. Tectonic influence on development of the Permian-Triassic unconformity and basal Triassic strata, Paradox basin, southeastern Utah. In *Mesozoic systems of the Rocky Mountain region, USA,* ed. M. V. Caputo, J. A. Peterson, and K. J. Franczyk, pp. 109–31. Denver: Rocky Mountain Section, Society for Sedimentary Geology.

Huntoon, P. W. 1993. Influence of inherited Precambrian basement structure on the localization and form of Laramide monoclines, Grand Canyon, Arizona. In *Laramide basement deformation in the Rocky Mountain foreland of the western United States,* ed. C. J. Schmidt, R. B. Chase, and E. A. Erslev, pp. 243–56. Special Paper 280. Boulder: Geological Society of America.

Lucchitta, I. 1972. Early history of the Colorado River in the Basin and Range Province: *Geological Society of America Bulletin* 83:1933–48.

———. 1989. *Overflight of the tectonic boundary between the Colorado Plateau and Basin and Range Provinces.* 28th International Geological Congress Field Trip Guidebook T116/389. Washington, D.C.: American Geophysical Union. 21 pp.

McKee, E. D., and E. H. McKee. 1972. Pliocene uplift of the Grand Canyon region— Time of drainage adjustment. *Geological Society of America Bulletin* 83:1923–32.

McKee, E. D., R. F. Wilson, W. J. Breed, and C. S. Breed, eds. 1967. *Evolution of the Colorado River in Arizona.* Bulletin 44. Flagstaff: Museum of Northern Arizona. 68 pp.

Nelson, S. T., J. P. Davidson, and K. R. Sullivan. 1992. New age determinations of central Colorado Plateau laccoliths, Utah: Recognizing disturbed K-Ar systematics and re-evaluating tectonomagmatic relationships. *Geological Society of America Bulletin* 104:1547–60.

Nelson, S. T., and D. G. Tingey. 1997. Time-transgressive and extension-related basaltic volcanism in southwest Utah and vicinity. *Geological Society of America Bulletin* 109:1249–65.

Peterson, F. 1994. Sand dunes, sabkhas, streams, and shallow seas: Jurassic paleogeography in the southern part of the Western Interior basin. In *Mesozoic systems of the Rocky Mountain region, USA,* ed. M. V. Caputo, J. A. Peterson, and K. J. Franczyk, pp. 233–72. Denver: Rocky Mountain Section, Society for Sedimentary Geology.

Peterson, F., and C. Turner-Peterson. 1989. *Geology of the Colorado Plateau.* Field Trip Guidebook T130. Washington, D.C.: American Geophysical Union. 65 pp.

Powell, J. W. 1873. Geological structure of district of country lying to the north of the Grand Canyon of the Colorado. *American Journal of Science* (3rd series) 5:456–65.

Spencer, J. E. 1996. Uplift of the Colorado Plateau due to lithosphere attenuation during Laramide low-angle subduction. *Journal of Geophysical Research* 101, no. B6:13595–609.

7

Quaternary Period

The Quaternary Period, which brings us to modern times, is subdivided into two epochs. The older of the two is the Pleistocene, which ranges from about 1.6 Ma to 10,000 years before present (B.P.). The younger Holocene Epoch extends from 10,000 years B.P. to the present. The Quaternary takes us from the realm of geologic time measured in millions of years into measurements of hundreds or tens of thousands of years.

The Pleistocene, commonly known as the Ice Age, was marked by multiple episodes of glaciation, periods in which much of the earth's land surface was blanketed by vast ice sheets. These great glaciers advanced and retreated over approximately 100,000-year cycles. During glacial periods, when huge volumes of water were stored as glacial ice, sea level dropped as much as 300 feet, exposing normally submerged areas such as the Bering Land Bridge, which linked the North American and Eurasian continents. During the interglacial periods, when the earth warmed, the glaciers retreated toward the polar regions and sea level rose, isolating the great land-masses.

Although the massive ice cap of the northern hemisphere never extended southward onto the Colorado Plateau, small isolated glaciers periodically draped higher areas such as the Aquarius Plateau and some of the widely separated mountain ranges. Regionally, these lower latitudes of North America were indirectly affected by the northern ice sheets. Effects include a cooler climate, related vertical shifts in vegetation zones, and the development of huge freshwater lakes that were fed by the glacial meltwater. Besides being cooler, the Colorado Plateau also was wetter. The ice sheet to the north caused the jet stream to split, sending moist air and increased precipitation southward onto the Colorado Plateau.

While the modern vegetation of the Colorado Plateau has not changed appreciably from that of the Pleistocene, the cooler climate pushed the vegetation zones to lower elevations. For example, the areas covered by extensive pygmy forests of piñon and juniper that are so familiar over the Plateau today were characterized by Colorado blue spruce, Douglas fir, and limber pine about 12,000 years ago. Curiously, fossil evidence from this time suggests that piñon pines did not exist on the colder, wetter Pleistocene Plateau and have only spread over the region since the termination of the last glacial episode 10,000 years ago.

Probably the most remarkable aspect of the Pleistocene Colorado Plateau was the assortment of large distinctive mammals, or *megafauna*, that roamed the open grasslands. These included the elephantlike mammoths and mastodons, enormous ground sloths, tapirs, musk oxen, bison, and horses. Many of these large mammals went extinct at the end of the Pleistocene. Some, however, including the grizzly and black bears, the bison, mountain lions, and wolves, survived and thrived into the Late Holocene. Unfortunately, increased human settlement of the region over the last 200 years resulted in the grizzly bear, wolf, and buffalo being hunted to extinction on the Colorado Plateau.

Major Normal Faults in Southwest Utah ≈

The southwest margin of the Colorado Plateau in Utah is cut by three major north-trending normal faults that gradually ease the otherwise undeformed Plateau into the intensely faulted and folded Basin and Range to the west (fig. 7.1). These down-to-the-west faults mark the deliberate migration of active crustal extension onto the Plateau. This region, called the High Plateaus by geologist C. E. Dutton (1880), serves as a transition between the Colorado Plateau and the Basin and Range. From east to west the lengthy faults are the Paunsaugunt fault, the Sevier fault, and the Hurricane fault (figs. 7.1 and 7.2). Although these faults probably initiated during the Late Tertiary, only Quaternary movement can be demonstrated with certainty; thus they are discussed in this chapter.

Paunsaugunt Fault ≈
The easternmost of these, the Paunsaugunt fault, separates the Tropic valley to the east from the scalloped east buttress of the Paunsaugunt Plateau to the west. This fault passes through the heart of Bryce Canyon National Park, which is perched along the southeast flank of the Paunsaugunt Plateau. In fact, Bryce Canyon owes much of its existence to this fault. The luminous hoodoos and fins of Bryce are carved into the pastel limestone and siltstone of the Claron Formation, which is fortuitously preserved on the downthrown west side of the fault. One seemingly contradictory feature of the upthrown east side is the topographically *low* Tropic valley, while the downthrown west side is *high*. The modern landscape is best understood if approached in terms of the relative resistances of the rocks involved.

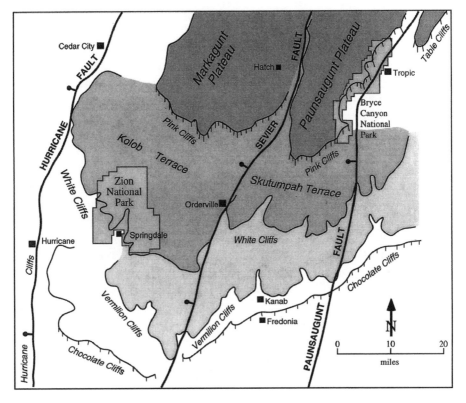

Fig. 7.1. Map of the High Plateaus region in southwest Utah showing the large down-to-the-west normal faults that form the transition zone between the Colorado Plateau to the east and the Basin and Range to the west. The region west of the Hurricane Cliffs is in the Basin and Range physiographic and geologic province.

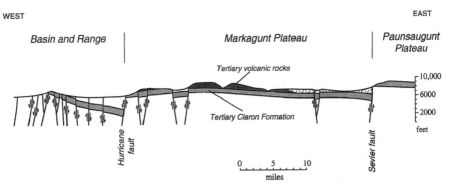

Fig. 7.2. Cross section across the Paunsaugunt and Markagunt Plateaus showing the Sevier and Hurricane faults as well as the Basin and Range area west of the Hurricane fault zone. The cross section is oriented west–east and located about 8 miles north of the town of Panguitch. The Claron Formation is shown as a reference point to illustrate the nature of the various faults in the region. After Best and Hamblin 1978.

When fault movement was initiated, the Claron blanketed the region. As the east side was pushed up into a high platform, the agents of erosion attacked it with a greater ferocity, removing the resistant carapace of Claron limestone to reveal a soft underbelly of Cretaceous sandstone and shale. Meanwhile the topographically lower Claron on the west side sat passively, bypassed by this erosional phase. As the incipient drainage network began to cut into the easily eroded Cretaceous sediments to the east, the high Claron to the west remained unaffected except for some lateral erosion into the limestone platform as the Paria River widened the Tropic valley. It was the lateral erosion and gullying along the edge of Paunsaugunt Plateau that formed the fantastic Bryce landscape. While faulting is the prime factor in molding Bryce Canyon, its bizarre, crenulated forms owe much to contrasting hardness between the Cretaceous and Tertiary strata that were juxtaposed by the crustal movements.

The Paunsaugunt fault has an exposed length of about 120 miles and stretches south toward the Utah-Arizona border, where it appears to die out (fig. 7.1). However, a short distance into Arizona, in line with the fault, lies the west flank of the Kaibab Plateau, which is defined by a system of similar north-trending normal faults. It is likely that these faults are a southern continuation of the Paunsaugunt. North of Bryce Canyon the amount of offset appears to diminish so that its surficial expression turns into a west-dipping monocline, presumably underlain by the fault. Vertical offset along the fault where Highway 12 crosses it in Bryce Canyon Park is about 1,200 feet. Here the pink Claron Formation on the west side has been dropped against the gray and brown Cretaceous rocks to the east, making recognition of the fault trace straightforward.

While the age of the Paunsaugunt fault has been speculated to be Late Tertiary, steep cliffs adjacent to the fault trace and low scarps in Holocene deposits (soil and alluvium) strongly support recent movement. Earthquakes recorded in Tropic and Panguitch in the early 1900s may have been generated by movement on the Paunsaugunt or Sevier faults.

Sevier Fault ≈
The Sevier fault bounds the Paunsaugunt Plateau on its west side and forms the Sevier River valley (figs. 7.1 and 7.2). The fault accommodates another step down to the west as we approach the edge of the Colorado Plateau. The fault is lengthy, stretching for 200 miles to reach far southward into Arizona, where it cuts across the Grand Canyon. Recent estimates based on seismic imaging of the subsurface in the Red Canyon area suggest up to 3,000 feet of down-to-the-west offset, although it may be as little as 100

feet in other places along the fault. The main pulse of movement on the fault apparently occurred between 7.6 and 5.4 Ma, in the Late Tertiary. This interpretation is based on dated volcanic rocks that have been affected to various degrees by fault movement. Essentially, the 7.6 Ma rocks show considerable offset across the fault, while 5.4 Ma rocks in the same area exhibit only minor offset. It should be pointed out that major faults of this length will show variable amounts of offset along different segments. Where datable, timing of movement will likewise show variability along its length. The fault continues to be active, as demonstrated by the offset of a 0.5 Ma basalt flow at Red Canyon, where Highway 12 crosses the fault.

Hurricane Fault ≈

The Hurricane fault is commonly used to draw the west margin of the Colorado Plateau in southwest Utah. The 160-mile-long fault runs from the Grand Canyon northward past Cedar City, Utah, where it forms the precipitous west boundary of the Markagunt Plateau (figs. 7.1 and 7.2). To the west lies the fault-riddled Basin and Range Province, where the crust of western North America is presently being pulled apart in an east-west direction. Over most of its reach the fault forms a steep, west-facing escarpment called the Hurricane Cliffs. The steep angle of this continuous cliff indicates that it has not yet submitted to erosional degradation, suggesting that at least some of the fault movement is young.

As in the case of the previously discussed faults, the time of initiation for the Hurricane fault cannot directly be determined. Some scientists suggest a Miocene age, while others argue for a younger Pliocene or Pleistocene origin. As we have seen in other long faults, total offset is highly variable along its extent. Near the town of Toquerville, just west of Zion National Park, the west side is believed to have dropped more than 8,000 feet, a vertical offset of one and a half miles! This, of course, occurred in increments. About 1,500 feet of this has taken place since the eruption of a Quaternary basalt flow across the fault. If Quaternary movement rates are calculated and extrapolated back in time to account for the total offset on the fault, a Late Miocene or Early Pliocene (about 5 Ma) fault initiation is suggested.

Evidence for recent movement on the Hurricane fault abounds. Detailed mapping along part of the fault by geologists Meg Stewart and Wanda Taylor (Stewart, Taylor, and others 1997) has revealed offsets of up to 20 feet in unconsolidated gravel at the base of the fault. Moreover, recent seismic activity in southwest Utah such as the June 28–29, 1992, earthquake swarm and a magnitude 5.2 earthquake near St. George on September 2, 1992, is attributed to movement on the Hurricane fault.

This large, active fault is the greatest natural hazard facing the exploding population of southwest Utah today and should be carefully monitored for signs of activity.

Glaciers in the Desert ≈

Pleistocene glaciation in southern Utah apparently was limited to one area, although elsewhere on the Colorado Plateau widely scattered highlands show the effects of flowing ice. The Aquarius Plateau, which lies west of Capitol Reef at an elevation of more than 11,000 feet, was blanketed by an extensive icecap in the latest part of the Pleistocene. Similar glaciated areas in surrounding parts of the region include the volcanic San Francisco Mountains that loom majestically above Flagstaff, Arizona. The La Sal Mountains that overlook Moab, Utah, likewise were subjected to a thick cover of glacial ice. Surprisingly, the over 11,000-foot Henry Mountains contain no evidence for the presence of glaciers. This absence may be the consequence of a Pleistocene rain shadow. The Henry Mountains lie immediately east of the High Plateaus area of the Colorado Plateau, which includes the Aquarius Plateau. As clouds were driven east across the Basin and Range by the jetstream, they gathered moisture. Upon encountering the loftier west edge of the Colorado Plateau they were pushed upward into cooler air. As the clouds became overloaded with moisture, precipitation in the form of snow fell onto the High Plateaus. By the time the clouds had drifted over the Henrys the moisture was exhausted, leaving the mountains high and dry.

Aquarius Plateau ≈

During the Late Pleistocene the top of the Aquarius Plateau, with an area of 70 square miles, was periodically covered by a thick icecap. The glaciers inched downslope from the center of the plateau to flow in a radial pattern, in a manner observed today in large continental icecaps in the northern latitudes. Evidence for a radial flow pattern can be seen in features carved into the hard basalt caprock, such as fluting, grooves, and striations. Fluting with several feet of relief formed as the great weight of the thick glacial ice, coupled with horizontal movement of the glacier, carved elongate furrows into the bedrock. Grooves and striations originated when the glacier plucked pieces of loose bedrock from the underlying surface and incorporated them into the basal ice. These rocks were then scraped along the bedrock as the ice moved, leaving gouges of varying depths as proof of their passage. All these linear features are oriented parallel to glacial move-

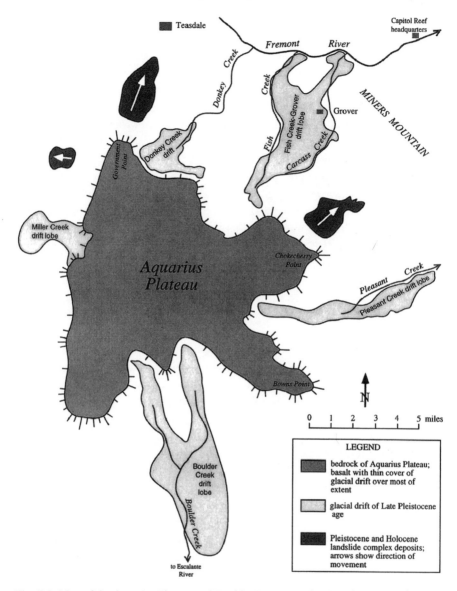

Fig. 7.3. Map of the Aquarius Plateau and Boulder Mountain showing the extent of Late Pleistocene glacial deposits and landslides. After Flint and Denny 1958.

ment and provide vital information on the direction of flow. On the Aquarius Plateau all of these indicators show radial movement, from the center to the edges of the plateau.

Additional evidence for glaciation on the Aquarius Plateau lies along its flanks, in the valleys that extend far below its summit. In contrast to the erosional features on the plateau top, these features are depositional, in the form of till—great piles of unsorted detritus that accumulated at the terminus of the glaciers. Distinctive lobes of till can be recognized in Fish Creek on the northeast side of the plateau and Pleasant Creek, which drains eastward; both drainages feed the Fremont River (fig. 7.3). Till is also present in the upper reaches of Boulder Creek, which flows south into the Escalante River. This sediment, ranging in size from clay to boulders, was bulldozed down these steep valleys by large tongues of ice that spilled over the plateau rim when the icecap was most robust. These icy fingers descended as low as 6,600 feet. At maximum extent the Fish Creek glacier extended far downvalley, past the present-day village of Grover, to the Fremont River. Glacial outwash, composed of fluvial sediment transported by glacial meltwater, was flushed far down the Fremont, past Capitol Reef.

Although the glacial drift on and around the Aquarius Plateau has not been directly dated, it has been speculated to range in age from more than 70,000 years to about 12,000 years B.P., when the last large glaciers in North America began to shrink. It is clear that several episodes of glaciation affected the area, based on relations between different drift deposits. Many of these deposits have been attributed to the youngest event at about 12,000 years B.P., simply because these should be the best preserved. Recently, new techniques in dating more distal glacial outwash deposits have been used to refine the glacial chronology of the region. This evidence lies in the river sediments of the Fremont River, far downstream from glacial activity.

Fremont River Terraces ≈

The basis of this new glacial chronology lies in the terraces composed of older river gravels that sit at several levels above the Fremont (photo 7). These terrace gravels range from about 100 to less than 20 feet thick. The ubiquitous black, rounded boulders that rim the Fremont River high and low were derived from the top of the Aquarius Plateau and are a conspicuous component of these terraces. According to research by James Repka and his colleagues (1997) at the University of California, Santa Cruz, these terraces record the waxing and waning of glaciers atop the plateau. By dating these gravels with innovative new isotopic chemical techniques, we can more accurately decipher the timing of discrete glacial events.

The sediment that makes up the terraces is interpreted to be glacial outwash, generated by the grinding action of the glaciers against bedrock. During glacial periods the huge volumes of outwash inundated the Fremont with sediment, filling it rapidly. Subsequently the glaciers receded

Photo 7. Flat-lying Quaternary terrace gravels atop the tilted, thinly bedded Jurassic Summerville Formation near the Bullfrog–Notom road in Capitol Reef National Park.

or even disappeared, and the sediment supply was shut off. It was at this time that the thick sediment fill was incised by the sediment-starved river, leaving stranded remnants of outwash preserved high above the canyon floor. These cycles of glacial deposition and interglacial erosion occurred several times in the Pleistocene, each time leaving a terrace as a record of the fluctuating conditions.

Repka and his colleagues used isotopes to date quartzite clasts collected from the Fremont terraces. Quartzite is a quartz-rich sandstone held together with quartz cement, producing an extremely resistant rock that is 100% quartz. Isotopes for which the rocks were analyzed were ^{10}Be (beryllium), which forms from oxygen, and ^{26}Al, which derives from silicon. Conveniently, oxygen and silicon are the sole constituents of the mineral quartz (SiO_2). The isotopes ^{10}Be and ^{26}Al are formed by the impacts of secondary particles produced by cosmic rays as they bombard the earth's surface. The analysis of clasts for these isotopes essentially is a measurement of the time that a clast was exposed to these rays at the surface. After the isotope measurements are obtained, a series of complex experiments and calculations is required to obtain the age of the terraces. After much experimentation, the results of Repka and colleagues appear to be well founded. However, the further refinement of their methods and the testing of their techniques by others ultimately will determine the accuracy of their dates.

Dates from the three most widely preserved terraces are 60,000 ± 16,000 years, 102,000 ± 16,000 years, and 151,000 ± 24,000 years B.P. While these terrace ages overlap with known glacial maximums determined by other methods, none record the last glacial interval more than 12,000 years ago, contrary to previous interpretations. It is possible that one of the less prominent, undated terraces represents this final glacial event, but positive identification awaits further study. If these dates for terrace construction and, by analogy, glaciation on the Aquarius Plateau do hold up to scrutiny, multiple episodes of glacial/interglacial activity occurred over a much longer interval than was previously supposed. They further suggest that the Pleistocene climate in southern Utah was a more dynamic and complex system than earlier evidence had indicated.

Quaternary Life ≈

The Arrival of Humans ≈

The best evidence for the appearance of humans (*Homo sapiens*) on the Plateau indicates that they arrived about 12,000 years B.P. It is believed that these people migrated southward into the region after earlier crossing the Bering Land Bridge that connected western Alaska with Asia. The land bridge evolved as a direct result of the widespread glaciation. As more and more water became locked up in the vast glaciers, sea level lowered enough so that the sea floor beneath the narrow Bering Strait became exposed, enabling humans, and also numerous species of animals, to cross into North America.

Two distinct cultures can be recognized in the artifacts of Pleistocene humans. The *Clovis* people (mammoth hunters) are the older of the two. The *Folsom* people (bison hunters) probably descended from the Clovis. While no remains of either of these people have been discovered on the Plateau, they can be differentiated by their distinctive stone spear points that were used for hunting. Clovis points and stone tools have been discovered in association with mammoth kill sites in North America, and we can be certain that Clovis people killed and fed on the great Pleistocene mammals, although direct evidence for this has yet to be discovered on the Colorado Plateau. Several intriguing petroglyphs that are interpreted to represent mammoths have been found along canyon walls of the Plateau, however, suggesting a similar association in this region.

Pleistocene Megafauna ≈

The Colorado Plateau has yielded a rich assemblage of Late Pleistocene fossil material that is particularly important in reconstructing the paleoenvironment for that time. Evidence collected thus far provides information on a fascinating megafauna of now-extinct mammals. Although the actual bones of these mammals are sparse, their dried dung or excrement is preserved in the many deep, dry sandstone alcoves that sit above the modern canyon floors. Surprisingly, these dung deposits contain a treasure trove of evidence for the mammals and the vegetation that made up their diet. The dung, as well as wood, bone, and hair, can be dated using ^{14}C (carbon), a radioactive isotope that is found in all organic material. This method of dating has its limits, however, as it is accurate only for ages less than 40,000 years; it is ideal for the late Pleistocene and Holocene. Work on these deposits of the Colorado Plateau and adjacent areas in the last twenty years has mostly been spearheaded by geologists Larry Agenbroad and Jim Mead (1987, 1989), who run the Quaternary Studies Program at Northern Arizona University in Flagstaff.

The most exotic fossil mammals so far discovered on the Colorado Plateau are the elephantlike mammoths and mastodons, the giant ground sloth, musk and shrub oxen, and camels. Other mammals include the bison, several species of ancestral horses, and three species of bear. Several carnivores survived the latest Pleistocene extinction, notably the gray wolf and the mountain lion, although the gray wolf has recently been hunted to extinction in the region. The secretive mountain lion, however, survives in the remote canyons and mesa tops today.

Pleistocene horses evolved from three-toed ancestors that are first recognized from the Early Tertiary (about 55 Ma) fossil record of North America, where they are believed to have originated. In the Late Pleistocene these mammals, which by then closely resembled the modern horse, became extinct on this continent. Their European descendants, however, survived this extinction and were later domesticated by humans. The European horses originated in North America and are believed to have migrated earlier to the Eurasian landmass across the Bering Land Bridge. In a quirky turn of events the horse was reintroduced in recent times to North America by European explorers.

The giant short-faced bear was the largest of the Pleistocene predators on the Colorado Plateau, but it also became extinct at the end of this icy epoch. Remains of these large carnivores have been discovered at Keams Canyon, on the Hopi Indian Reservation in northern Arizona. These bears migrated to North America from South America, across the narrow Central American land bridge. Remains of the grizzly and black bear have also been recognized in Pleistocene deposits on the Plateau. In contrast to the giant short-faced bear, these species originated in Eurasia and migrated across the Bering Land Bridge sometime during the Pleistocene. Both of these bears survive in North America today, but since the widespread human settlement in the last 200 years the grizzly has been hunted to extinction on the Colorado Plateau.

According to Agenbroad and Mead (1989), forty-one localities containing mammoth remains have been discovered on the Colorado Plateau (fig. 7.4). Most of these occur in the Colorado River canyon and its tributaries, including recently discovered skeletal remains from a tributary of the lower Escalante River. Besides giving us an appreciation of the abundance of mammoths on the Colorado Plateau, these skeletal remains and dung deposits provide much-needed insight into the timing of their extinction, as well as the vegetation they fed on and the environment in which they apparently thrived.

The deep sandstone grottos whose darkened entrances pock the Navajo and Wingate sandstones in the canyons of the Colorado River system are ideal localities for mammoth remains. One of these, Bechan Cave in Glen Canyon National Recreation Area, contains what is probably the largest mammoth dung deposit in North America. The name "Bechan" is Navajo for large excrement. Individual dung balls, called boluses, range up to about 25 cm in diameter and form a blanket on the cave floor, beneath a layer of more recent Holocene sand and silt. Dung is composed mostly of grass, sedge, and rushes. While it appears that Bechan was used as a mammoth toilet, the deposits indicate use by other members of the community as well. Dung of the Shasta ground sloth, shrub ox, Harrington's mountain goat, and a species of horse, all long extinct, has also been identified from the dung blanket. Associated hair preserved in the dry cave environment has been identified for the mammoth, two species of sloth, shrub ox, horse, bison, and bear. [14]C dating of mammoth and sloth dung in Bechan Cave has so far yielded dates from 13,505 ± 580 to 11,630 ± 150 years B.P.

Dung and skeletal material from mammoths have been dated by the ^{14}C method throughout the Colorado Plateau. Agenbroad and Mead report thirty-eight dates from various studies that reach as old as 30,800 ± 1700 years B.P. The majority of these (twenty-one) fall between 11,000 and 13,550 years, when the final glacial event was waning. The youngest of these ages was from a locality in Professor Valley, at the base of the La Sal Mountains. Here a calculated average of several dates from the same locality produced a date of 11,270 ± 65 years B.P. When this is compared with minimum ages obtained from other megafauna on the Plateau, we see consistently similar dates. The youngest age for the Shasta ground sloth comes

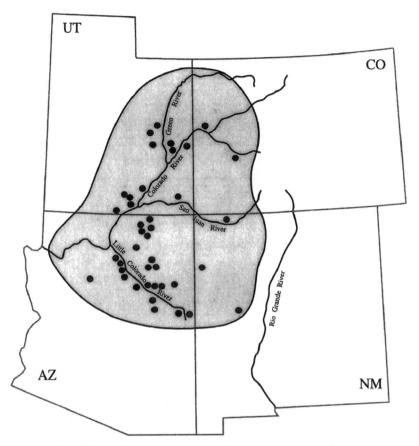

Fig. 7.4. Map of the Colorado Plateau (shaded) showing known localities of mammoth remains. Black circles are localities with mammoth bones, while shaded circles designate localities for dung or other remains. After Agenbroad and Mead 1989.

from the Grand Canyon, where a date of 11,018 ± 50 years was obtained (Martin and others 1985). Likewise, the average age for thirty-seven analyses from horn sheaths and dung of Harrington's mountain goat, also from the Grand Canyon, is 11,160 ± 125 years B.P. (Mead and others 1986). Taken collectively, these dates show an amazingly consistent age for megafauna extinction on the Colorado Plateau.

The trigger for Late Pleistocene megafauna extinction on the Colorado Plateau has long been the topic of debate. Some have attributed this event to the arrival of humans on the scene, while others have called on a change in climate. Evidence for the overlapping of humans and large mammals in North America is clear in the form of mammoth kill sites. A new and more intelligent predator such as *Homo sapiens* may have stressed the ecosystem to such a level that it eventually collapsed. The demise of the giant herbivores would in turn affect the larger carnivorous predators by reducing their food supply. This scenario, however, is somewhat difficult to accept, considering the wide extent of mammals and the relative paucity of Paleoindian remains on the Plateau. Possibly the timing was coincidental.

The currently favored hypothesis for the extinction of this mammalian fauna is a warming climate that provoked a shift in vegetation zones across the North American continent. The problem with this hypothesis is the recognition of alternating warm and cool climates throughout the Pleistocene. Why did these mammals survive the earlier warming trends? It is possible that the combination of a change in vegetation (food) and the arrival of humans was too much for the large mammals to endure. Evidence supporting a dramatic change from Late Pleistocene to Holocene plant communities is abundant.

Late Pleistocene Flora ≈

The record of Late Pleistocene flora of southern Utah has been well preserved in several ways, thanks to the dry Holocene climate. Recognizable plant material has been recovered from mammoth dung blankets just below the surface of large sandstone caves and alcoves and packrat middens nestled high on protected ledges, as well as in the terraces of sandy alluvium exposed along the canyon floors.

Packrats, small rodents of the genus *Neotoma*, serve as unexpected allies in scientists' attempts to decipher the Late Quaternary history of southwestern North America. The dwellings of these widespread rodents are constructed of the most detailed record of Late Pleistocene/Holocene environments available. The dwellings grow as the packrat collects plant and other materials that are piled into the den. This is an ongoing process. When built in areas sheltered from water, such as rock alcoves and caves, the dens become "cemented" by packrat urine to form a resistant, brown-black, tar-like mass of organic material called a *midden*.

These packrat "collections" provide an unparalleled window into the changing environment over thousands of years. These dwellings, for the most part, are continuously occupied; when one packrat dies, another quickly takes over the den, continuing the collection and accumulation process. Materials built into these middens include anything within up to 150 feet of the den, so the record is both local and long-term. The material mostly consists of plant macrofossils—leaves, flowers, and stems of the surrounding vegetation that the packrat fed on. Any insect or vertebrate remains within its range also are incorporated into the dwelling. Pollen blown in by the wind sticks to the tarry urine as it would to flypaper, providing a more extensive record of the regional flora. Finally, middens provide a chronology of these changes because they are datable by the ^{14}C method. This creates the potential for a detailed floral record over the last 40,000 years, the upper limit for ^{14}C dating.

In 1990 Julio Betancourt of the U.S. Geological Survey published a superb study of packrat middens from a cave situated at 7,300 feet on the south flank of the Abajo Mountains in southeast Utah. This study combined plant macrofossil assemblages with ^{14}C dating to obtain a detailed chronology of Late Pleistocene/Holocene vegetation changes. These data point to an increasingly warm and arid climate. Betancourt found that prior to 11,000 years B.P. the area was dominated by Engelmann spruce, subalpine fir, and limber pine (fig. 7.5). Today Engelmann spruce and subalpine fir at these same latitudes are found above about 9,200 feet, more than 2,000 feet higher than their Late Pleistocene range (fig. 7.6). Accompanying the decline of these species, the Douglas fir and Ponderosa pine appeared on the scene and spread rapidly. The juniper also appeared about this time, but its expansion was gradual. Piñon pines did not arrive until about 5,000 years B.P. and, with the juniper, blanket the region today (fig. 7.5).

In a compilation of data from packrat middens from northern Mexico to the Colorado Plateau Thomas Van Devender (1986) of the Arizona–Sonora Desert Museum in Tucson tracked the distribution of the piñon pine geographically and through time (fig. 7.7). These data reveal that before about 11,000 years B.P. the piñon was abundant far to the south, in Mexico and south Texas, down to an elevation of about 2,000 feet. It became extinct or declined rapidly to extinction in the region after 11,000 years. There is no record of the piñon anywhere on the Colorado Plateau before this time; it is first recognized between 11,000 and 10,000 years in the Grand Canyon area. From there it apparently spread slowly across the Plateau region. Today the piñon inhabits an elevation range of 5,300 to 7,000 feet in southern Utah.

Studies of mammoth dung from side canyons that are tributary to the Colorado and Escalante Rivers indicate that the huge mammals fed largely on the grasses that dominated the setting. While their diet consisted predominantly of grasses, locally it was augmented with reeds that grew along the ponded parts of canyon floors. Woody species provided variety and included sagebrush, saltbush, birch, rose, and cactus. Collectively this assemblage of plants from more than 12,000 years ago suggests a sagebrush-steppe upland type of setting. Streamside communities included a forest of birch, elderberry, dogwood, and spruce. Today these canyons and their surrounding mesas are characterized instead by blackbrush desert scrub. The nearest possible analogue for the Pleistocene flora that has been recognized

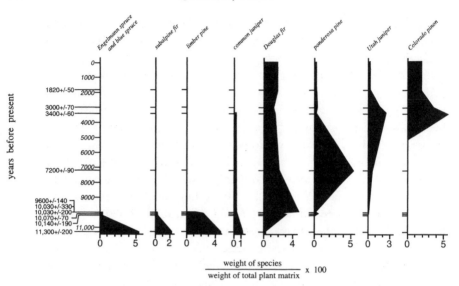

Fig. 7.5. Sequence of selected flora from Allen Canyon Cave on the south flanks of the Abajo Mountains in southeast Utah. Data come from macrofossils in packrat middens. Specific dates come from ^{14}C analyses of midden material. Note the decrease at 11,000 years of spruce, subalpine fir, and limber pine, which is concurrent with an increase in Douglas fir. Also shown is the later appearance of ponderosa pine and Utah juniper and the much later appearance of the piñon pine. After Betancourt 1990.

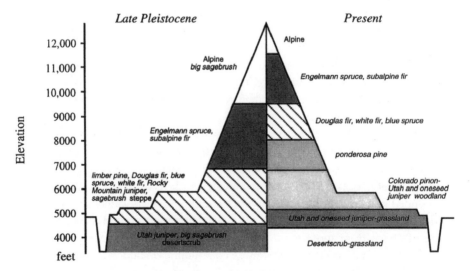

Fig. 7.6. Generalized vegetation zones for the Late Pleistocene (about 14,000–12,000 years B.P.) and the present on the Colorado Plateau. After Betancourt 1990.

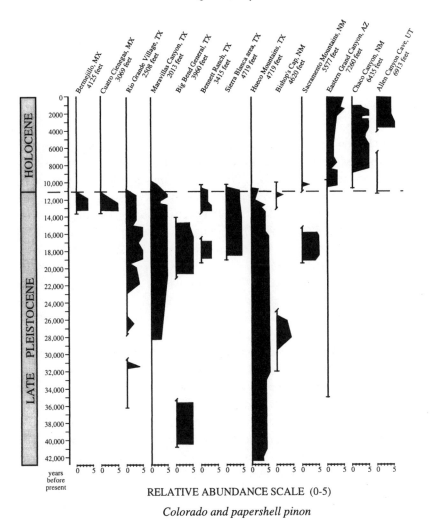

RELATIVE ABUNDANCE SCALE (0-5)

Colorado and papershell pinon

Fig. 7.7. Relative abundances of Colorado and papershell piñon pines during the Late Pleistocene and Holocene. Data were obtained from packrat midden sequences along a latitudinal transect south from Durango, Mexico, to southeast Utah. The length of the baseline indicates the time range of midden coverage in each area. Relative abundance scale: 1 = rare, 2 = uncommon, 3 = common, 4 = very common, 5 = abundant. After Van Devender 1986.

lies on distant adjacent highlands of the Henry Mountains and the Aquarius Plateau, nearly 4,000 feet above the canyons in which the mammoth dung was discovered.

The End of the Pleistocene ≈

The vegetation record from numerous sites, and several types of deposits across the Colorado Plateau, shows a pronounced shift at about 11,000 years B.P. toward a modern flora. Prior to this time, during the glacial interval, most plant associations were depressed with respect to elevation: Pleistocene plants were found at much lower elevations than their modern counterparts (fig. 7.6). For instance, Betancourt recognized a pre–11,000 year association of spruce, fir, and limber pine at 7,300 feet. Today these species are either gone from the region (limber pine) or limited to a zone above 9,200 feet. The absence of piñon on the Plateau during this time suggests that even the lowest elevation was too cool for it.

Similar relationships have been found in the lower Escalante River drainage at an elevation from 3,600 to 4,300 feet. Kim Withers and Jim Mead (1993) found that pre–11,000 year terrace sediments contained remnants of spruce, fir, and mountain mahogany. They also found abundant evidence for well-watered species such as wild rose and water birch. Later records (< 11,000 years B.P.) indicate an absence of these species from the area. Spruce-fir forests presently are confined to the upper southern slopes and top of the adjacent Aquarius Plateau, above 8,900 feet. Mountain mahogany is found today more than 800 feet above these canyons, and water birch about 300 feet above. Finally, the Pleistocene treeline was lowered by about 2,000 feet, expanding the range of alpine tundra.

It is generally believed that the Late Pleistocene climate was cooler and wetter than today's climate, resulting in the depressed vegetation zones and apparently producing an ideal environment for the mammals that roamed the region and browsed on the plants. Differences in temperature and precipitation did not produce simple beltlike zones based on elevations, but also were influenced by local factors such as adjacent highlands and waterways. Perennial streams in deep canyons continued to provide habitat for some of these Pleistocene plant associations well into the Holocene.

Debate among Quaternary geologists becomes heated when turning to the question of *how much* cooler and wetter the Late Pleistocene climate was. Some scientists argue, based on the high water level of Pleistocene lakes to the west, that rainfall was considerably higher, but that temperature had not necessarily decreased much. An increase in precipitation could not have been uniform, and estimates range from 60 to 100% more than seen today in southwestern North America. Still others argue for a 7° to 11° C cooler temperature on the Colorado Plateau, but no more precipitation than today. Withers and Mead (1993) estimated a temperature of 3° to 4° C cooler for the lower Escalante canyons based on plant fossils, but attributed the wetter conditions to a higher water table. While there is no consensus, most scientists attribute the depressed elevation zones to a cooler and wetter climate in the Late Pleistocene. The question remains as to how much cooler and wetter.

Extinction of the Late Pleistocene megafauna and the shift of vegetation to increasingly higher elevations on the Colorado Plateau were coincident with a warmer, drier climate. The death of the larger mammals probably is linked to the change in their food supply, which was hastened by the worldwide climate change. But warmer climates had prevailed during previous interglacial periods; why did this one result in widespread extinctions? A growing body of evidence suggests that this warming event, which continues today, was more pronounced and longer-lived than previous ones. The shift in vegetation zones, which in some cases takes several thousand years, seems to support this idea; so does a continuation of this interglacial period today. A problem with comparing earlier interglacial periods with the present one arises when trying to date associated deposits. The ^{14}C dating method is accurate only back to 40,000 years B.P., leaving the previous 1,660,000 years of the Pleistocene without a simple dating technique for the commonly preserved organic materials. Like many problems of the geologic past, the Late Pleistocene megafauna extinction remains to be solved. While it seems that we constantly are inching toward an answer, every one brings up more questions. Such is the nature of science.

The Holocene Epoch ≈

The Pleistocene/Holocene boundary is a wandering one and has been placed anywhere from 12,000 to 6,000 years B.P., depending on whom one listens to and how it is defined. Some scientists have placed the boundary at the last rapid worldwide rise of sea level. This rise was triggered by the melting of the continental ice caps, which in turn was caused by an increasingly warmer climate. Each of these events has also been used to define the boundary. Widespread glaciation was over by about 12,000 years, but sea level may not have peaked until about 8,000–6,000 years B.P. Thus the problem becomes apparent. On the Colorado Plateau, which was not directly affected by glaciation or sea-level changes, the shift in vegetation zones and megafauna extinction took place about 11,000 years ago. This is probably as good a boundary as any, at least for our purposes.

The Holocene, also known as the Recent, extends to the present and represents the modern environment and climate. It also encompasses the substantial and continuing effect of humans on the landscape. In southern Utah this effect locally is great (e.g., Lake Powell), but in large areas it is minimal, which is what makes the region so appealing.

Geologically, not a lot has happened on the Colorado Plateau in the Holocene. A contributing factor is the short time span of about 11,000 years. In geologic time this is an instant; most geologic processes require longer periods to become evident. For example, movement along the faults that mark the tectonically active west margin of the Plateau probably has occurred in this time span, but it has not dramatically altered the landscape. Erosion of highlands has continued, as it always will in the presence of water, but even this has slowed considerably because of the increasingly arid Holocene climate.

Escalante River Basin ≈

Holocene river deposits in the deep canyons of the Escalante River and its tributaries have been the target of several studies. The purpose of these investigations was to unravel the Holocene flood history of the basin and possibly determine the causes for flood events. As we have seen, throughout geologic history the most helpful information is stored in the sedimentary record; the Holocene is certainly no different. Large floods locally left deposits up to 10 feet thick. Flooding and deposition was followed by long-term erosion and incision into the sediment by normal flow, leaving the thick accumulations of sand and silt perched high above the active channel. The ages of many flood-related terraces have been determined from ^{14}C dating of twigs and wood buried in the flood debris and from hearths or burned horizons on the terrace surface, the product of Anasazi Indian occupation.

In a study of terraces by Paul Boison and Peter Patton (1985, 1986), the Holocene record of sedimentation and erosion was deciphered for Harris Wash, Twentyfive Mile Wash, and Coyote Gulch, all western tributaries to the lower Escalante River. Surprisingly, these studies suggest that the Early Holocene history of the three adjacent drainage systems differs markedly. In fact, no correlations between recognizable Early Holocene events could be made. Variations in terrace deposits between the canyons may be attributed to differences in the extent of individual drainage networks, and possibly to landslide activity along the adjacent Straight Cliffs. Easily eroded Cretaceous rocks of the Straight Cliffs form the heads of these canyons; in-

creased mass wasting activity along this escarpment could contribute excessive amounts of sediment downstream, resulting in an isolated but large-
scale influx of sediments.

One notable outcome of this study is the correlation of historic flood deposits marked by the partial burial of large cottonwood trees in these
canyons. Dendrochronological or tree-ring dating studies of these trees
document an increase in flooding over the past 100 to 150 years. Similar
timing has been established in the Pine Creek drainage, which runs off the
southern flanks of the Aquarius Plateau. Flood-scarred ponderosa pine trees
indicate that the magnitude and frequency of flooding increased after 1880
and increased further after 1909. Correlative historic floods throughout the
Escalante basin correspond with the settlement of the town of Escalante
and the attendant grazing of large herds of sheep and cattle and timber-
cutting on the slopes of the Aquarius Plateau.

While prehistoric floods in various drainages of the Escalante River appear to have no connection, the recognition of region-wide historic flooding immediately following settlement strongly suggests a human element in
these recent events. The increased flood frequency likely is tied to changes
in land use that stripped large parts of the area of vegetation. In fact, a correlation between the degradation of rangeland and arroyo cutting and
human settlement and grazing has been made across southwestern North
America.

The Modern Colorado Plateau of Southern Utah—A Rant ≈

Although much of southern Utah is protected as National Parks and
Monuments, human-induced degradation of the environment continues.
For example, cattle still graze and trample large parts of Capitol Reef and
Grand Staircase–Escalante Monument even though numerous studies have
documented damage to the land. It is ironic that visitor centers warn hikers
to avoid stepping on fragile cryptobiotic soil even as cattle stomp it into
oblivion. Additionally, cattle in the arid backcountry congregate around
springs and seeps, driving away the native animals that rely on these water
sources for survival. Too many times on the Colorado Plateau I have come
upon a much-anticipated spring only to find it fouled with cow excrement
and the delicate vegetation flattened.

Logging on the lofty Aquarius Plateau also threatens these "protected"
lands. This heavily forested tableland serves as the headwaters for the
Escalante River and the Fremont/Dirty Devil system, two of the largest
rivers in southern Utah. Clearcut logging, besides serving as an impetus for

bulldozing new roads into pristine forest, accelerates soil erosion, which results in the washing of huge quantities of silt and mud into the streams and rivers. Thus the quality of precious river water declines.

Finally, construction of Glen Canyon dam has been the ultimate insult to the fantastic Colorado Plateau landscape. Beneath the oily, flotsam-covered monotony of Lake Powell lies the most spectacular canyon system on the Plateau, Glen Canyon. The few remaining remnants of what once existed are present in the tributaries to the lower Escalante River. It is a heart-wrenching experience to hike down one of these fantastic canyons, with anticipation growing as the canyon walls grow higher and the canyon floor narrows, only to round a bend and be confronted with the skeletons of cottonwood trees jutting from the stagnant water of Lake Powell. It somehow reduces one's hope for humankind to imagine that such a sacred place could be defiled by human hands.

But hope is not lost. While humans have desecrated portions of this great land, the damage is not irreparable, and large tracts of wild country still exist. When humans are removed from the equation, as they most certainly will be eventually, time will heal the ravaged grasses, and trees will once again grow tall on the sites of massive clearcuts. The dam will weaken and collapse, and the scars of Lake Powell will fade. Yet this is no reason to abandon the call for preservation and an end to the degradation. Much of the damage cannot be repaired in our lifetime. But what about our children and their children? It would be a tragedy for them to grow up with less wild country to inspire them than what we now enjoy. For the sake of future generations we must fight for the preservation of this wonderful wild country of southern Utah.

References

Agenbroad, L. D., and J. I. Mead. 1987. Late Pleistocene alluvium and megafauna dung deposits of the central Colorado Plateau. In *Geologic diversity of Arizona and its margins: Excursions to choice areas*, ed. G. H. Davis, and E. M. Vanden Dolder, pp. 68–84. Special Paper 5. Tucson: Arizona Bureau of Geology and Mineral Technology.

———. 1989. Quaternary geochronology and distribution of *Mammuthus* on the Colorado Plateau. *Geology* 17:861–64.

Best, M. G., and W. K. Hamblin. 1978. Origin of the northern Basin and Range province: Implications from the geology of its eastern boundary. In *Cenozoic tectonics and regional geophysics of the western Cordillera*, ed. R. B. Smith and G. P. Eaton, pp. 313–40. Memoir 152. Boulder: Geological Society of America.

Best, M. G., E. H. McKee, and P. E. Damon. 1980. Space-time-composition patterns of late Cenozoic mafic volcanism, southwestern Utah and adjoining areas. *American Journal of Science* 280:1035–50.

Betancourt, J. L. 1990. Late Quaternary biogeography of the Colorado Plateau. In *Packrat middens: The last 40,000 years of biotic change*, ed. J. L. Betancourt, T. R. Van Devender, and P. S. Martin, pp. 259–92. Tucson: University of Arizona Press.

Boison, P. J., and P. C. Patton. 1985. Sediment storage and terrace formation in the Coyote Gulch basin, southcentral Utah. *Geology* 13:31–34.

Dutton, C. E. 1880. *Geology of the High Plateaus of Utah*. Washington, D.C.: Dept. of Interior, U.S. Geographical and Geological Survey of the Rocky Mountain Region. 307 pp.

Elias, S. A. 1997. *The Ice-Age history of southwestern National Parks*. Washington, D.C.: Smithsonian Institution Press. 200 pp.

Flint, R. F., and C. S. Denny. 1958. *Quaternary geology of Boulder Mountain, Aquarius Plateau, Utah*. Bulletin 1061–D. Washington, D.C.: United States Geological Survey. 164 pp.

Graf, W. L., R. Hereford, J. Laity, and R. A. Young. 1987. Colorado Plateau. In *Geomorphic systems of North America*, ed. W. L. Graf, pp. 259–302. Centennial, special vol. 2. Boulder: Geological Society of America.

Kottlowski, F. E., M. E. Cooley, and R. V. Ruhe. 1965. Quaternary geology of the southwest. In *The Quaternary of the United States*, ed. H. E. Wright and D. G. Frey, pp. 287–98. Princeton: Princeton University Press.

Lundin, E. R. 1989. Thrusting of the Claron Formation, the Bryce Canyon region, Utah. *Geological Society of America Bulletin* 101:1038–50.

Martin, P. S. 1987. Late Quaternary extinctions: The promise of TAMS 14C dating. In *Nuclear instruments and methods in physics research*, B29, pp. 179–186. Amsterdam: North-Holland.

Martin, P. S., R. S. Thompson, and A. Long. 1985. Shasta ground sloth extinction: A test of the blitzkrieg model. In *Environments and extinctions: Man in late glacial North America*, ed. J. I. Mead and D. J. Meltzer, pp. 5–14. Orono: University of Maine, Center for the Study of Early Man.

Mead, J. I., L. D. Agenbroad, A. M. Phillips III, and L. T. Middleton. 1987. Extinct mountain goat (*Oreamnos harringtoni*) in southeastern Utah. *Quaternary Research* 27:323–31.

Mead, J. I., P. S. Martin, R. C. Euler, A. Long, A. J. T. Jull, L. J. Toolin, D. J. Donahue, and T. W. Linick. 1986. Extinction of Harrington's mountain goat. *National Academy of Science Proceedings* 83:836–39.

Nelson, L. 1990. *Ice Age mammals of the Colorado Plateau*. Flagstaff: Northern Arizona University Press. 24 pp.

Patton, P. C., N. Biggar, C. D. Condit, M. L. Gillam, D. W. Love, M. N. Machette, L. Mayer, R. B. Morrison, and J. N. Rosholt. 1991. Quaternary geology of the

Colorado Plateau. In *Quaternary nonglacial geology: Conterminous U.S.*, ed. R. B. Morrison, pp. 373–406. The Geology of North America, vol. K-2. Boulder: Geological Society of America.

Patton, P. C., and P. J. Boison. 1986. Processes and rates of formation of Holocene alluvial terraces in Harris Wash, Escalante River basin, south-central Utah. *Geological Society of America Bulletin* 97:369–78.

Repka, J. L., R. S. Anderson, G. S. Dick, and R. C. Finkel. 1997. Dating the Fremont River terraces. In *Geological Society of America Field Trip Guide Book, 1997 annual meeting, Salt Lake City*, ed. P. K. Link and B. J. Kowallis, pp. 398–402. Geology Studies, vol. 42, part 2. Provo: Brigham Young University.

Rowley, P. D., T. A. Steven, and H. H. Mehnert. 1981. Origin and structural implications of upper Miocene rhyolites in Kingston Canyon, Piute County, Utah. *Geological Society of America Bulletin*, part 1, 92:590–602.

Shroder, J. F., and R. E. Sewell. 1985. Mass movement in the La Sal Mountains, Utah. In *Contributions to the Quaternary geology of the Colorado Plateau*, ed. G. E. Christenson, C. G. Oviatt, J. F. Shroder, and R. E. Sewell, pp. 48–85. Special Studies 65. Salt Lake City: Utah Geological and Mineral Survey.

Stewart, M. E., W. J. Taylor, P. A. Pearthree, B. J. Solomon, and H. A. Hurlow. 1997. Neotectonics, fault segmentation, and seismic hazards along the Hurricane fault in Utah and Arizona: An overview of environmental factors in an actively extending region. In *Geological Society of America Field Trip Guide Book, 1997 annual meeting, Salt Lake City*, ed. P. K. Link and B. J. Kowallis, pp. 235–52. Geology Studies, vol. 42, part 2. Provo: Brigham Young University.

Van Devender, T. R. 1986. Late Quaternary history of pinyon-juniper-oak woodlands dominated by *Pinus remota* and *Pinus edulis*. In *Proceedings of pinyon-juniper conference*, ed. R. L. Everett, pp. 99–103. U.S. Department of Agriculture, Forest Service Technical Report Int-215.

Van Devender, T. R., R. S. Thompson, and J. L. Betancourt. 1987. Vegetation history of the deserts of southwestern North America: The nature and timing of the Late Wisconsin–Holocene transition. In *North America and adjacent oceans during the last deglaciation*, ed. W. F. Ruddiman and H. E. Wright, Jr., pp. 323–52. The Geology of North America, vol. K-3. Boulder: Geological Society of America.

Withers, K., and J. I. Mead. 1993. Late Quaternary vegetation and climate in the Escalante River basin on the central Colorado Plateau. *Great Basin Naturalist* 53:145–61.

Plate 1. Stevens Arch in the lower Escalante River drainage is carved from the red rocks of the Glen Canyon Group. The lower ledgy rocks are composed of the Kayenta Formation, an amalgamation of lens-shaped Jurassic river channels. The upper cliffs through which the arch is cut are the widespread Navajo Sandstone, part of a Jurassic dune field that blanketed the region about 200 million years ago.

Plate 2. The Castle in Capitol Reef and the Lower Mesozoic rocks of the Waterpocket Fold. Rock units from bottom to top include the thinly bedded red Triassic Moenkopi Formation, colorful shales of the Triassic Chinle Formation, and the broken cliffs of the Castle, composed of the Triassic/Jurassic Wingate Sandstone. The white cliffs on the skyline are the Jurassic Navajo Sandstone.

Plate 3. Colorful limestone and siltstone of the Tertiary Claron Formation make up the crenulated pinnacles of Bryce Canyon National Park.

Plate 4. Colorful shales of the Upper Jurassic Brushy Basin Member of the Morrison Formation. Shales are floodplain deposits of rivers that traversed the region. Dinosaur bones are a common component of these rocks.

Plate 5. Gray Tropic Shale and yellow Ferron Sandstone make up the base of the thick Cretaceous succession along the east side of the Waterpocket Fold. Laccoliths of the Henry Mountains loom in the background.

Plate 6. Factory Butte, a prominent landmark just west of Hanksville, is hewn from gray shales of the Blue Gate Shale and is capped by the Emery Sandstone. These rocks were deposited on the floor of the Cretaceous Western Interior seaway, which alternated between deep (gray shale) and relatively shallow (sandstone) water.

Plate 7. Towers of west Zion National Park. The lower thin cliff band is the Jurassic Kayenta Formation; the Jurassic Navajo Sandstone makes up the upper red and white cliffs.

Plate 8. The Escalante monocline looking north up Pine Creek. The Jurassic Carmel Formation dips west to plunge into the subsurface beneath the town of Escalante. The Navajo Sandstone makes up the white cliffs at the crest of the monocline.

Plate 9. Looking north along the Sevier fault on the west edge of the Paunsaugunt Plateau. The Sevier fault is a normal fault with the relatively young black basalt dropped down on the left (west side) relative to the older Tertiary Claron Formation on the right (east side).

Plate 10. The Cockscomb, a landmark near the town of Teasdale, is composed of shattered Navajo Sandstone that has been tilted along a fault to form a resistant ridge.

Plate 11. Looking east into the monoclinal fold of Capitol Reef with the Henry Mountains in the background.

Plate 12. Temple of the Sun in Cathedral Valley, in the remote northern part of Capitol Reef National Park, is carved from tidal-flat and sand-dune deposits of the Jurassic Entrada Formation.

Plate 13. Jurassic rocks of the Escalante River drainage. Domes of Navajo Sandstone are underlain by ledgy sandstone of the Kayenta Formation.

II≈

Road Logs:
The Geology
of Southern Utah

8

Hanksville to Capitol Reef National Park
Visitor Center via State Highway 24

Highway 24, from the town of Hanksville west to the Capitol Reef Visitor Center, follows the upstream course of the Fremont River. It is the scouring action of its water and sediment over millions of years that has carved this passageway. The Fremont River canyon is one of the few east-west routes through the rugged Waterpocket Fold, a regional north-south–trending spine of resistant rock. If not for the river and its canyon, the spectacularly folded rocks of Capitol Reef would indeed be a formidable obstruction to travel.

Immediately downstream (east) from Hanksville the Fremont merges with Muddy Creek, which cuts eastward through the San Rafael Swell, and the two become the Dirty Devil River (fig. 8.1). The Dirty Devil cuts through deep red canyons toward its rendezvous with the Colorado River (now Lake Powell). Its splendid name harkens back to John Wesley Powell's second Colorado River expedition in 1872, when the Dirty Devil's water ran thick and brown as the result of large amounts of suspended sediment. Downstream, later in the trip, the clear, sweet waters of Bright Angel Creek in the Grand Canyon were named as a counterpart to the Dirty Devil.

Heading upstream along the Fremont, we encounter a variety of sedimentary rock types, some very resistant to erosion, others not so resistant. More resistant rocks such as sandstone and conglomerate tend to form deep canyons with high, steep walls. In contrast, easily eroded rocks like shale and siltstone are seen as slopes in broad, open valleys. Keeping these characteristics in mind, we can correlate changes in the nature of the valley with changes in rock type.

As we drive west along the Fremont River, numerous rock formations are beautifully displayed on the surrounding walls and slopes, each unit representing a shift in the environment from the previous one. These strata range in age from Early Triassic (about 245 million years ago) to Late Cretaceous (about 85 m.y. ago), covering most of the Mesozoic Era, also known as the age of dinosaurs. In Hanksville we begin in Jurassic-age

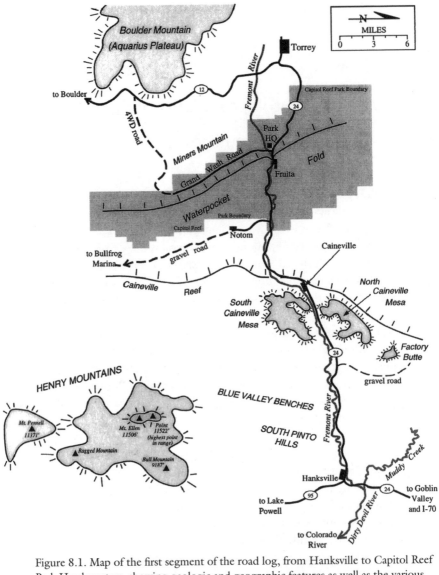

Figure 8.1. Map of the first segment of the road log, from Hanksville to Capitol Reef Park Headquarters, showing geologic and geographic features as well as the various towns, settlements, and roads. The route follows Utah Highway 24 west along the Fremont River for most of the way, except where it cuts through the Caineville Reef.

rocks, formed during the middle of the Mesozoic, and journey upward through increasingly younger rocks until we peak in the Late Cretaceous near the settlement of Caineville (see figs. 8.1 and 8.2). West of Caineville the strata become sharply tilted to the east as we enter the Waterpocket Fold, a regional north-south–trending uplift that has been incised by the river. From this point the narrowing canyon cuts through increasingly older and colorful rocks to emerge in Fruita, the site of the Capitol Reef visitor center, where Early Triassic rocks form the foundation of a spectacular encirclement of steep, multicolored walls. In brief, the rocks we pass through initially become younger and then older, so that many of the formations will be seen a second time, but in reverse order (fig. 8.2).

Hanksville ≈

Hanksville initially was called Groves Valley, for Walter H. Groves, who joined geologist G. K. Gilbert during his fieldwork in the nearby Henry Mountains. Groves mapped the topography of the region, while Gilbert concentrated on the surrounding geology. In 1885 the name was changed to Hanksville for Ebenezer Hanks, who settled this area with his family in 1882. Unlike many western towns, Hanksville has never experienced a boom, although its population peaked at around 500 in the early 1970s, during the waning days of the uranium rush on the Colorado Plateau. Today Hanksville is a quiet town, its main street lined with motels, gas stations, and restaurants catering to people heading east to Lake Powell or west to Capitol Reef. I hope it will escape the fate of many of today's "quaint" western towns and remain small and quiet.

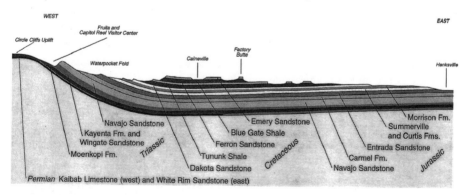

Figure 8.2. Cross section showing the geology from Hanksville to Capitol Reef Visitor Center along Utah Highway 24. The figure is schematic, with no vertical or horizontal scale implied.

Henry Mountains ≈

The Henry Mountains can always be seen looming to the south of Hanksville, hazy purple-blue in the summer and fall or white with snow in the winter and spring. The range was named by John Wesley Powell for Professor Joseph Henry, a physicist and ardent supporter of Powell's exploratory trips. Henry was the first secretary of the Smithsonian Institution and was critical in raising much-needed funds for Powell's Colorado River expeditions.

Remarkably, the Henrys were the last mountain range in America to be placed on a published map. The area remained a mystery until Powell sighted them on his 1869 expedition down the Colorado River. It was only after an arduous climb from the river to the rim of Glen Canyon that he was able to describe them and map them accurately. It is easy to see why it took so long to explore these mountains: they are completely encircled by some of the most treacherous and rugged country in North America, which, of course, is much of its attraction today.

It was not until 1872 that Anglos set foot on the slopes of the Henry Mountains, when A. H. Thompson and a group of men were sent by Powell to recover a boat that was left behind at the Dirty Devil River earlier in the expedition. Before embarking on his 1871–72 trip, Powell had left instructions for a pack train to meet them at the mouth of the Dirty Devil River so they could be resupplied. So little was known of the region at this time that the packers mistook the Escalante River for the Dirty Devil and waited there instead. Eventually Powell's group abandoned the Dirty Devil but left a boat behind, only to find the supplies downstream at the mouth of the Escalante. The group wintered at Lee's Ferry with plans to continue the following spring. During this time Thompson was sent by Powell to recover the boat and meet them at Lee's Ferry. Thompson had additional instructions to investigate the Henry Mountains en route. His group departed from the Mormon settlement of Kanab, eventually to emerge high on the east slope of Boulder Mountain. From this vantage point the maze of rocks separating them from the Henrys was clearly laid out. Upon viewing this terrain, Thompson stated, "It is cut with deep canyons and looks impassable." Still, it was their duty to explore this unknown country, and Thompson's men were excited at the prospect. Eventually they found a faint Indian trail that led eastward, taking them across the barren slickrock fins of the Waterpocket Fold, through the mesas of Cretaceous strata, and finally to the foot of the isolated mountain range. Upon reaching springs at the base of the mountains, the group split up: some climbed Mt. Pennell, while the others ascended Mt. Ellen, the

highest point in the range at 11,522 feet. Mt. Ellen was named for
Thompson's wife, who also was Powell's sister; Mt. Pennell is believed to
have been named for a relative or close friend of Thompson's.

The first geological investigation of the Henry Mountains was
conducted soon after this exploratory trip, in 1875 and 1876, by preemi-
nent geologist Grove Karl Gilbert. Powell's earlier views of this isolated
range had piqued his interest. He arranged for Gilbert, who had joined
the Powell Survey in 1875, to travel to the mountains and determine their
origin. The region was still a vast, unmapped wilderness; like Thompson's
party before them, Gilbert's party relied on dim prehistoric Indian trails to
lead them to waterholes and springs. Gilbert spent two weeks in the
Henrys in 1875 and returned for two months the following year. His field
studies culminated in his classic book published in 1877, *Report on the
Geology of the Henry Mountains.* Gilbert's early work on the region was of
such high quality that distinguished geologist Charles B. Hunt, who studied the Henry Mountains from 1935 to 1939, found little that disagreed
with Gilbert's work. Hunt's work was published as USGS Professional
Paper 228 (Hunt and others 1953), entitled *The Geology and Geography of
the Henry Mountains.*

Gilbert's studies showed that the Henry Mountains formed from igneous intrusions rather than volcanic eruptions, as was previously
supposed. Igneous intrusions are created when rising molten rock or
magma does not quite force its way to the earth's surface. Instead, the
magma comes to rest somewhere below and cools slowly. This slow cooling rate allows large mineral grains or crystals to grow, producing a coarse-
grained plutonic igneous rock.

Plutonic rock bodies come in a multitude of sizes and geometries. The
Henry Mountains are composed of several unique mushroom-shaped bodies that Gilbert termed laccoliths. Gilbert first recognized this type of intrusion in the Henry Mountains and described them in detail in his 1877
report. This report has insured that the mountains, though isolated even
today, are known throughout the geological community as the premier
example of laccoliths.

Laccoliths are discrete, mountain-sized blisters of magma that intrude
sedimentary rock layers at shallow levels in the crust. As the magma slowly
rises through the sedimentary sequence, it pushes laterally, exploring for
weaknesses between the layers. Upon discovering a defect, the viscous
mass pushes its way into the vulnerable spot, prying the layers apart. As it
forces its way laterally, the magma body domes the overlying strata
upward, forming the characteristic mushroom-shaped intrusion (fig. 8.3).
Long after the magma has cooled and crystallized, erosion dissects the

A

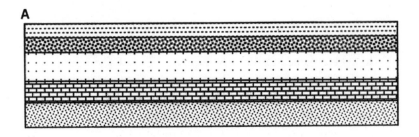

B

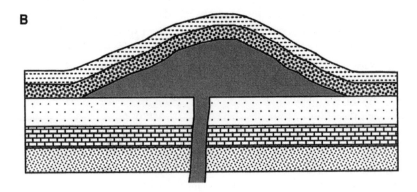

C

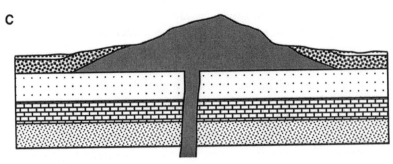

Figure 8.3. Evolution of laccolith-type igneous intrusions that make up the Henry Mountains, Navajo Mountain, the La Sal Mountains, and the Abajo Mountains on the Colorado Plateau. The exceptionally well-exposed rocks in the Henry Mountains were responsible for the idea of laccoliths, first proposed by eminent geologist G. K. Gilbert: blisters of rising magma squeeze between sedimentary layers, conforming to the top of the underlying layer and doming up the overlying layers. *A*: flat-lying sedimentary rocks prior to being intruded by magma. *B*: mushroom-shaped laccolith intrusion (shaded) squeezed between the previously flat-lying sedimentary succession. Note the flat bottom of the laccolith that conforms to the underlying sedimentary unit and the domed overlying units. The stem of the laccolithic "mushroom" is the conduit that fed magma to the main body. *C*: exposed plutonic core of the laccolith after the less-resistant overlying sedimentary layers have been stripped away by millions of years of erosion. In reality, the Henry Mountains consist of several overlapping laccoliths and are more complex.

uparched strata, eventually to reveal the granitic core. Most of the laccoliths in the Henry Mountains cluster have matured to this stage today, leaving the bare granite exposed and fringed by an encirclement of flatirons composed of baked sedimentary rock.

According to chemical analyses reported by geologist Stephen Nelson at UCLA and his colleagues in a 1992 report, the Henry Mountains laccoliths were intruded about 25 to 30 million years ago, during Late Oligocene time. These ages were obtained by using the latest technology for analyzing single mineral grains from the plutonic rocks for isotopes of the element argon, contained within the crystals. This age range makes the Henrys about the same age as similar laccoliths in the nearby La Sal Mountains and the Abajo Mountains, both easily seen to the east from the Henry Mountains.

Surprisingly, the Henrys host the only free-roaming buffalo herd in the lower forty-eight states. These buffalo are the descendants of a small pioneer herd of 18 that was transplanted from Yellowstone National Park in 1941. Today the herd has multiplied to about 200 animals. The present population has split into several small herds, which can often be seen grazing on the green upper slopes of the mountains in the summer and on the lower mesas and benches in winter.

The Jurassic Journey ≈

mile

0.0 ≈   On the way in and out of Hanksville, regardless of the route, the orange and red striped cliffs of Jurassic Entrada Sandstone cannot be missed. Here the Entrada consists of thin maroon layers of siltstone and mudstone coupled with thicker beds of orange sandstone. Through time this originally continuous sheet of rock has been dissected into a system of fantastic fins, towers, and hoodoos. North of Hanksville, along Highway 24, remnants of the Entrada can be seen in a sandy basin in the form of widely separated slender pinnacles. Goblin Valley State Park is located in this same area and consists of a red Entrada amphitheater occupied by an army of squat Entrada goblins.

If we had stood here 175 million years ago, we would have seen to the north a vast, featureless mudflat, intermittently covered by a shallow sea; to the south, the sensuously curved hills of shifting sand—the border of a regional sand sea or erg that stretched south, east, and west for many miles. Periodically the area was fringed by sand dunes slowly migrating over the desiccated mudflats; at other times, during violent storms or high tide, the sea washed in to mix mud and silt with the sand. In this way the alternating rock types slowly accumulated, layer by layer. During Entrada deposition this part of southern Utah was a transition zone between two very different environments, a battleground between sand dunes and shallow sea, dominated by change.

1.2 ≈   West of Hanksville the crumbly green and white cliffs of the Curtis Formation bound the north banks of the Fremont River as we cross the bridge. The colors in the Curtis contrast with the red rocks of the Entrada Sandstone below and Summerville Formation above. The sandstone and limestone of the Curtis reflect a shallow sea and a regional geography that differed little from that of previous Entrada deposition.

1.4 ≈   The Summerville Formation is the next attraction on our Jurassic tour. The formation is easily recognized by its thin, continuous beds of orange, brown, and white. These pin-striped rocks mark the return of an extensive tidal flat. The rock that makes up the thin white beds and litters the rubble mounds at the base of the cliffs is gypsum, a soft evaporite mineral. The gypsum was left behind when mineral-rich seawater evaporated from low-lying pools that dotted the Summerville tidal flat. As we approach the strange fluted columns of Summerville Formation, a web of irregular white veins becomes apparent in the rock. These veins, while also composed of gypsum, have a different origin than the white beds that are part of the horizontal layering. These discordant veins precipitated after the formation had been deposited, from mineral-rich groundwaters that passed slowly through fractures in the buried rock.

2.8 ≈   Heading west, up through increasingly younger rocks, we pass through the inconspicuous Tidwell Member of the Morrison Formation. The Tidwell weathers into greenish-white slopes and represents the concluding episode in a series of mudflat deposits. The dearth of good Tidwell outcrops is due to the presence of gypsum. Sometime during Tidwell deposition the sea receded northward out of southern Utah, leaving behind an expansive shallow basin of trapped mineral-rich water. As the hot sun baked the flats, the water slowly evaporated, concentrating the waters to a brine rich in calcium and sulfur. Gypsum gradually formed from this brine and accumulated along with mud and silt. Because any contact with water easily dissolves rock gypsum, even in the extremely arid climate of the Hanksville area, the gypsiferous sediments dissolve into powdery slopes when exposed at the surface.

3.1 ≈   Tidwell slopes are capped by the red-brown ledges of the Salt Wash Member of the Morrison Formation. Salt Wash sandstone and conglomerate represent the channels of ancient rivers. The deposits of these east-flowing rivers provide the first hint of tectonic upheaval to the west, in central Nevada. Upon close inspection (stop the car and get out—stretch those legs!), the smooth, water-worn pebbles embedded in these ledges attest to their long, abrasive journey from their mountainous source. Crossbedding is the most obvious feature of the flat rock faces, with its graceful curves and intricate geometric patterns reminiscent of a well-executed work of art. In reality they are the design of turbulent waters and chaotic currents.

3.6 ≈   As we motor uphill to the bench formed by the top of the Salt Wash, one of the most colorful units on the Colorado Plateau, the Brushy Basin Member of the Morrison, dominates the immediate surroundings. These soft mounds of fine-grained rock display a palette of colors, in limitless shades of red, purple, green, and gray. The soft curves of the mounds are occasionally broken by thin white limestone ledges, lending a slight angularity to the scene. Shale and mudstone were deposited on a vast floodplain that only sporadically was cut by east-flowing rivers. Mud and silt accumulated on the floodplain during exceptionally high flood events, when the sediment-laden rivers spilled over their banks. Limestone probably formed in small, low-lying areas where ponded water sat for some time.

Dinosaurs roamed the Morrison floodplains in abundance, some feeding on the vegetation that grew along the rivers, others preying on these oversized vegetarians. The Morrison Formation is probably the most famous source for dinosaur skeletons in the world. The famed dinosaur quarry at Dinosaur National Park is in the Morrison, as are the dinosaurs found around the Grand Junction, Colorado, area. As yet no major dinosaur discoveries have occurred in the Morrison in this area, but it is probably only a matter of time before someone discovers a piece of bone jutting from a slope, only to realize it is part of a larger skeleton. They are out there.

To the Depths of the Cretaceous Sea ≈

6.3 ≈   Passing through the upper Morrison, we depart the colorful Jurassic rocks to enter the gray, barren world of the Cretaceous Period. This abrupt change is shocking, as if the earth had suddenly been drained of color. At midday the sun-scorched scene can be forbidding. In the early morning or evening, however, the low light and deep shadows create an alluring monochromatic landscape. The rocks also are a study in contrasts. The deeply incised folds of the slopes seem to clash with the harsh angularity of the mesa tops. Blue-gray shale slopes dominate the immediate surroundings and are the petrified muds of a deep sea that once covered the region. The distant mesa tops are sandstone, snapshots of a time when the seas momentarily shifted eastward, to be replaced by beaches or, where rivers flooded into the sea, large deltas.

Cretaceous sedimentation in southern Utah was influenced by two major factors, a rising mountain belt to the west and the dominating presence of a vast seaway. These ancient mountains, which geologists call the Sevier orogenic belt, were situated in present-day Nevada and western Utah. This north-south–trending mountain belt, which reached south into Mexico and north into Canada, was the result of intense compression acting on the western margin of the continent. Compressive forces continuously folded and faulted these mountains through the Cretaceous Period, an interval of about 70 million years.

The Sevier mountains were bordered to the east by a wide sedimentary basin that extended as far east as present-day Kansas. Large parts of the basin periodically were inundated by an elongate sea called the Western Interior seaway. Fluctuations in the level of this sea ultimately controlled the nature of the Cretaceous rocks we see today.

The Dakota Formation is the basal Cretaceous unit in this part of the region and forms a 40-foot sandstone cliff that locally lines canyon walls. In detail, three units can be recognized in the Dakota, each representing a unique environment. The basal unit is a brown, pebbly sandstone laid down in east-flowing rivers that drained the Sevier orogenic belt. These rivers ran to a sea that, at this time, lay in Colorado. Up close, the sandstone and conglomerate–filled channels are very similar to those seen earlier in the Salt Wash Member of the Morrison. Sweeping crossbeds and large, rounded pebbles attest to the forces of ancient river currents. The middle of the Dakota consists of several feet of black, sandy coal, the remnants of a densely vegetated swamp. The swamp evolved as sea level climbed and the shoreline shifted westward, temporarily backing up the rivers to create a stagnant marsh. The top unit is composed of sandstone deposited in a shallow sea or beach setting and reflects the continued westward push of the shoreline. Within this unit are the shells of fossil oysters and the curious-looking bivalve *Gryphaea*, which resembles a gray, gnarled toenail. Most original sedimentary structures such as crossbedding have been obliterated by the burrowing activities of ancient organisms that thrived in this shallow sea. The organisms themselves are unknown, but traces of their passage cover the surfaces of many beds. These Cretaceous beach-dwellers tunneled through the sand in search of organic material to feed on.

6.8 ≈   As we emerge from the canyon onto the top of the Dakota, a broad valley comes into view. The valley is surrounded by silver-gray aprons of Tununk Shale. The flat mesa tops above are capped by the Ferron Sandstone. Black shale of the Tununk marks the continued rise in sea level that began during Dakota deposition. As the sea expanded, Dakota sands gave way to deep sea muds. The shoreline slowly shifted westward, drowning the alluvial plain in its path. At its climax the shoreline reached into western Utah, almost to the base of the Sevier mountain belt. The Tununk represents a time when the deep sea floor of southern Utah was being blanketed with a bottomless black ooze. The black color of the shale is caused by the inclusion of large amounts of organic material mixed with the mud. Normally, where oxygen is abundant, organic matter rapidly decomposes or is consumed by other organisms. The depths of this sea, however, were devoid of oxygen, and the black, carbon-rich organic matter was preserved. Although this harsh sea-floor environment hosted no living organisms, fossil ammonites can be found in the Tununk. Ammonites were "floaters": they lived and fed while floating in the water column. Upon death, they sank into the deep dank ooze, where they lay undisturbed, to be buried under thousands of feet of subsequent sediment.

8.2 ≈   Geologically speaking, the Tununk invasion of the sea was short-lived. The Ferron Sandstone that caps the shale is the product of a large delta, a system that stretched great fingers of sediment eastward into the deepening sea. As the sea paused for a moment, a pulse of sand and gravel from the adjacent Sevier mountains pushed the shoreline back east, depositing sand over the deep sea muds. Deltas lined the coast of the Western Interior sea and were fed huge quantities of sediment by the rivers that headed in the Sevier belt.

10.2 ≈   The road rises gradually through the Tununk, then it climbs rapidly through a small canyon in the Ferron. Upon entering this canyon, it becomes apparent that these sands did not enter the sea as one abrupt flood of sediment. The first indications of an advancing wave of sediment are thin sandstone beds in the midst of black shale, suggesting that initially small spurts of sand were jetted into the shallowing sea. These were followed by increasingly large inpourings of sediment, as evidenced by the gradual increase in sandstone thickness upsection. Near the top of the Ferron individual sandstone beds reach up to 20 feet thick. Up close, these beds exhibit a variety of structures such as crossbedding and horizontal laminations. Some units have been completely churned up by organisms that lived in the sediments, producing a structure known as *bioturbation* (photo 8). Bioturbation is common in shallow-marine deposits,

Photo 8. The feeding traces of Cretaceous organisms on a bedding surface of the Ferron Sandstone along Highway 24 between Hanksville and Caineville.

and several inferences can be drawn from its presence in the Ferron. First, its sudden appearance as the dominant feature suggests that organisms were attracted in large numbers to this new, unoccupied shallow-water environment to root for nourishment in the fresh sand and mud. Second, burrowing organisms are able to live in sediments only if the deposition rate is low. If large amounts of sediment are continuously streaming in, the organisms would rapidly be buried and die. Thus, the presence of bioturbation suggests that sediment inflow was sporadic: at times low so that organisms could feed in the sand, at other times high, burying the organisms and allowing some of the sedimentary structures in this new deposit to be preserved, at least until more organisms could worm their way into it.

At the top of the Ferron the road again flattens, following the top of the resistant sandstone. Initially we are surrounded by low black mounds of silty coal, the Ferron's dying gasp before the sea returned. Paper-thin shale laminae sandwiched in these impure coals contain the flattened remnants of plants—twigs, leaves, and stems preserved as filmlike carbonaceous impressions. Coal swamps formed as the sea returned to flood the delta, eventually drowning the entire region.

After we pass through the coaly mounds, the landscape opens into the flat expanse of the Blue Gate Shale. This soft rock has been molded by the rare passage of water across its surface. The mazelike network of rills and ridges defines the unique desolation of these badlands. Except for the sparse, small shrubs that dot the alkaline clay, this landscape is devoid of vegetation.

10.9 ≈   Soon after we enter the gray shale expanse, a well-maintained gravel road heads north, toward the landmark of Factory Butte and farther to the San Rafael Swell. Unlike most roads through Cretaceous rocks in the southwest, this one is quite good because it follows the bench of stable Ferron Sandstone rather than cutting across the shale. Cretaceous shale throughout the Four Corners region is infamous for its degradation into an impassable black gumbo when wet.

Factory Butte is a prominent landmark in the Henry basin. Its blue-gray slopes consist of 1,500 feet of deeply gullied Blue Gate Shale. The relatively flat top, from which the landform gets its name, is the Emery Sandstone. Erosion of the Emery is controlled by prominent vertical joints, probably formed by tectonic stresses associated with uplift of the nearby Waterpocket Fold and San Rafael Swell. The pattern of these vertical joints on Factory Butte gives the appearance of a large cluster of buildings bristling with smokestacks, mounted on a massive gray pedestal.

11.9 ≈   The Blue Gate Shale represents another westward invasion by the sea and a setting very similar to that of the earlier Tununk Shale. For tens of millions of years, particle by particle, mud accumulated on the still sea floor, attaining a thickness of more than 1,500 feet. The deep black color of the Blue Gate again points to the lack of oxygen in these quiet depths.

The succeeding Emery Sandstone marks another surge of sediment shed from the Sevier orogenic belt. Like the Ferron Sandstone before it, the Emery represents a return to shallow-water conditions as the influx of sediment pushed the sea back and allowed the land to reclaim its former territory.

13.2 ≈   On our drive west through the Blue Gate Shale, the flat tops of the massive North and South Caineville Mesas dominate the skyline. As we approach them, gray striated cliffs of shale loom menacingly over the north side of the road. It is rare that these Cretaceous shales form cliffs, but these are temporary features, geologically speaking. Where this once continuous mesa has been cleaved into lesser north and south segments by the Fremont River, the shale closes in from both sides, narrowing the valley considerably. Geologist G. K. Gilbert called this part of the valley Blue Gate, for the walls of unstable blue-gray shale that crowd the valley here. The formal rock unit, the Blue Gate Shale, was named for Gilbert's geographic reference to the area. On the south side of the

15.0 ≈ highway, bordering the Fremont River, are cultivated fields that form a patchwork of lush greenery in the spring and summer. Eventually the valley widens to accommodate the townsite of Caineville. This Mormon settlement, even in its heyday, scarcely qualified as a town, consisting of a schoolhouse and church that acted as a center for the pioneers who lived and farmed along the river and outlying areas.

Caineville ≈

15.1 ≈   Caineville was first settled in 1882 by Elijah Behunin and his family, who traveled from western Utah on the first recorded wagon passage through the Waterpocket Fold. A short time later, in February 1883, the community of Bluevalley, later named Giles, was founded a few miles downvalley by another group of Mormon settlers. Soon a system of irrigation canals and ditches was established up and down the Fremont valley. The low elevation, coupled with the constant supply of water from the Fremont River, promoted the growth of many different crops, including various fruits, melons, grains, most vegetables, and alfalfa for livestock.

For several years the Mormon settlements along the Fremont prospered, and several more small communities were founded. The Muddy and Fremont Rivers, whose life-sustaining waters flowed year round, allowed this barren land, with barely 5 inches of precipitation a year, to be productive. The irrigation systems that tapped these waters were simple affairs, with small dams and ditches diverting upstream water to be efficiently distributed to the fields.

In the fall of 1897 disaster struck the settlements when flooding teamed up with the geologic setting of the valley to destroy the water system. Before the flood the river channel was wide and shallow, making it easily accessible and simple to cross. Drinking water came from wells that extended 20 to 30 feet into the floodplain sediments. The flood, however, changed all this. The excess energy of the floodwaters caused rapid downcutting into the soft, easily eroded shale that floored the valley. Incision of 10 to 20 feet was common and over most of its course was accompanied by channel widening. Dams and irrigation ditches were destroyed or filled with sediment, and much of the irrigated cropland was buried beneath a thick layer of clay and silt. Every town in the valley was inundated, and half the town of Caineville was washed into the river (fig. 8.4).

The catastrophic erosion resulted in a wide, steepwalled channel that even today makes river access difficult. Besides destroying the entire system of irrigation ditches and dams, the dramatic drop of the channel floor

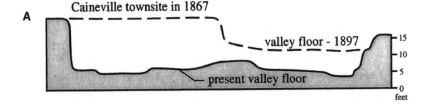

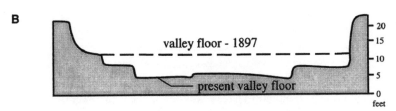

Figure 8.4. Profiles across the Fremont River valley showing the amount of erosion that has taken place since the first catastrophic flood in 1867. Profile *A* is at the old townsite of Caineville. Note the destruction of part of the actual townsite. Profile *B* is across the Fremont River at the old townsite of Bluevalley, approximately 10 miles downstream from Caineville. After Hunt and others 1953.

made the rebuilding complex and laborious. The lower river level also dropped the groundwater table, drying up many of the wells on the floodplain.

Many of the settlers fled the valley after the flood, although a few of the persistent ones braved the destruction and rebuilt the dams and ditches to accommodate the new river level. But the forces of nature endured, and repeated flooding in subsequent years continued to wreak havoc on the settlers' efforts. Finally, in 1909, the Mormon church released those who wished to leave from their commitment to settling the valley. Population in the valley dropped from 552 in 1893 to only 256 in 1910 (Hunt and others 1953). Giles and Caineville were practically abandoned. The population of Hanksville, which was less affected by the flood, decreased only slightly.

Waterpocket Fold ≈

18.1 ≈   As we pass by the old Caineville townsite, the well-preserved 100-year-old church/schoolhouse is seen to the south, while sharp, clifflike exposures of the Blue Gate Shale obstruct the northern view. To the west we are confronted with the east-tilted fin of Ferron Sandstone that constitutes the Caineville Reef (fig. 8.1). The gap in this otherwise continuous ridge was cut by Caineville Wash as it pushed southward to rendezvous with the Fremont River at the townsite of Caineville. This reef marks the beginning of the buckled and tilted strata of the Waterpocket Fold, the feature we will travel through for the remainder of our journey into Capitol Reef National Park. The Waterpocket Fold is a monocline, a type of large-scale geologic structure that can be seen throughout the Colorado Plateau. Most folds in layered rocks have two limbs that dip in opposite directions and are separated from one another by an axis. Monoclines, however, consist of a single limb that dips in one direction—hence their name. The San Rafael Swell immediately to the north is another one of these structures. Early pioneers on the Colorado Plateau called the resistant ridges formed by these structures "reefs" because they often acted as barriers to travel. Monoclines on the Colorado Plateau are the surface expression of intense compressional forces that produced faults deep beneath the surface. As rocks at depth were pushed over adjacent rocks by these forces, overlying sedimentary layers were passively folded, flexing into single-limbed folds. This compressional deformation was a part of the Laramide orogeny, a mountain-building event that followed closely on the heels of the earlier Sevier orogeny. The Laramide orogeny took place about 65 to 50 million years ago and was also responsible for the formation of the modern Rocky Mountains in New Mexico, Colorado, Wyoming, and Montana. The monoclines of the Colorado Plateau are a much different expression of this mountain-building event than are the Rocky Mountains. The difference is due to preexisting weaknesses in the earth's crust that were reactivated in the Rocky Mountains, compared to a relatively stable Colorado Plateau.

18.5 ≈   The Waterpocket Fold has many different expressions, depending on the resistance of a particular unit to erosion. Sandstone tends to form abrupt, narrow ridges called *hogbacks*. Shale and siltstone typically are easily incised to form confined drainages called *strike valleys*. Such valleys are parallel to and bounded by the hogbacks.

Escape from the Cretaceous Badlands ≈

19.0 ≈   As we pass through the Ferron hogback, a dirt road leads north to Cathedral Valley, a spectacular and seldom-visited part of Capitol Reef National Park. The gray strike valley we drop into has been cleaved into the Tununk Shale by Caineville Wash. The highway turns south to follow the soft substrate of the valley, enlisting its easily eroded ancient mud in the modern problem of road design. Bounding the valley to the west, we return to the colorful mounds of the Brushy Basin Member of the Morrison Formation. Crumbly scablike remnants of sandstone clinging to the sides of these pastel-colored shale hills are pieces of the Dakota Sandstone and Cedar Mountain Formation. The Cedar Mountain forms the base of the Cretaceous succession here and is the product of turbulent, east-flowing rivers. The Cedar Mountain has been preserved only locally in Cretaceous hollows and valleys and was absent where we passed through the Jurassic-Cretaceous boundary farther east.

21.6 ≈ For several miles the road follows the gentle curves of the Tununk strike valley, crossing the Fremont River where it cuts another gap through the Caineville Reef. Gradually the road curves west, and we begin our journey backward through time, traversing increasingly older strata and eventually encountering the oldest yet, those of the Triassic Period. Traveling westward, we cross the brilliantly colored Brushy Basin shales, a welcome change from the drab gray and brown Cretaceous rocks from which we emerge.

25.0 ≈ As we once again enter the drainage of the Fremont River, large black cannonballs appear to be spilling over all the hilltops. In reality, these cannonballs are well-rounded boulders and cobbles of basalt, a dense, black volcanic rock that caps Boulder Mountain, about 20 miles to the west (fig. 8.1), where these rocks originated. For as long as the Fremont has flowed these boulders have tumbled down its channel—their rough edges worn smooth and polished by thousands of years of flowing water. Their location far above today's riverbed attests to a long-ago time, before the river had cut down to its present level. As the river slowly sliced its way down through the bedrock of Waterpocket Fold, earlier river deposits were left stranded, forming terraces that delineate the former river levels. Multiple episodes of downcutting have left several different terrace levels at various heights above today's canyon floor.

The Fremont River floodplain is wide where the river dissects the Brushy Basin shale, the widening enhanced by the lateral forces of the river on the soft shale. The valley is shaded by the canopies of large cottonwood trees, taking advantage of the unfailing water supply of the river. Throughout the arid canyon country of southern Utah the welcome green of the cottonwood is a reliable marker of springs and permanent waterways.

26.0 ≈ Leaving the Brushy Basin, the road begins to meander as it is forced to follow the circuitous route of the deepening canyon. As we descend stratigraphically, Jurassic river channels of the Salt Wash Member line the canyon walls. Winding alongside the river the canyon again widens where the Tidwell Member and

27.6 ≈   the Summerville Formation intersect the canyon floor. Once again a broad floodplain borders the riverbanks—and once again weak rock is responsible. These tall cottonwoods mark the eastern boundary of Capitol Reef National Park.

27.9 ≈   Here also the Bullfrog–Notom road heads south from Highway 24 (fig. 8.1). This well-maintained gravel road follows strike valleys of Cretaceous and Jurassic strata as it parallels the Waterpocket Fold. It extends 65 miles southward to Bullfrog marina on Lake Powell, but ultimately circles the Henry Mountains to end up back in Hanksville. The stretch of road along the Waterpocket Fold is encompassed by incredible vistas. To the west lie the bare slickrock spines and domes of Navajo Sandstone. To the east are high cliffs of mesa-capping Cretaceous sandstone underlain by steep slopes of gray shale; behind this the Henry Mountains tower as a stately backdrop.

The settlement of Notom lies just 4 miles south of the junction with Highway 24, along the banks of Pleasant Creek (fig. 8.1). The small village was first settled in 1886 and today consists of a few orchards and houses set back from the road in a deep glade of cottonwoods.

Back on Highway 24, at the park boundary, is a rest area with large shade trees and a restroom. From this rest area, looking north to the canyon wall across the river, we get a clear view of several Jurassic rock units that are not usually well exposed. From top to bottom they are the brown sandstone and conglomerate lenses of the Salt Wash Member, the crinkly red, white, and green beds of the Tidwell Member, and just above the river the orange, red, and white pinstripes of the Summerville Formation. Looking back down the canyon, we can see the upper slopes of the Brushy Basin, capped by the black boulders of an earlier incarnation of the Fremont River.

28.1 ≈   Continuing west on Highway 24, we are immediately confronted with the Entrada Formation—orange cliffs of thin siltstone and thicker sandstone beds. Hidden somewhere beneath the slopes that separate the Entrada and the Summerville lie the gypsum-rich beds of the Curtis Formation.

29.0 ≈   A short distance upcanyon the Carmel Formation emerges from beneath the brick-red base of the Entrada. The Carmel is recognized by its green, gray, and red beds of limestone, shale, and gypsum. The lower part of the formation is folded and contorted, due to the willingness of gypsum to flow plastically when subjected to the pressure of overlying strata. The base of the Carmel is marked by deep red shales that overlie the massive white cliffs of Navajo Sandstone.

29.8 ≈   The canyon walls close in rapidly and the shadows deepen as we enter the Navajo Sandstone, the premier canyon-forming unit of Capitol Reef and the Colorado Plateau. The buff-colored cliffs are etched with the huge, sweeping crossbeds that are the hallmark of the Navajo. These crossbeds are the exhumed internal structures of monstrous sand dunes that blanketed a sand sea on a scale greater than anything seen today. While this great sand pile certainly took many millions of years to accumulate, the exact age of the Navajo is poorly known. The aridity of the vast sand sea was too harsh for all but the most extreme creatures, making age-diagnostic fossils extremely rare. We do know, however, that the underlying Kayenta Formation is of Jurassic age, as is the overlying Carmel Formation.

The Navajo Sandstone is the uppermost of the three formations of the Glen Canyon Group: the Wingate Sandstone, the Kayenta Formation, and the Navajo. The group was named for its unparalleled exposure in Glen Canyon, before it was drowned beneath Lake Powell. These three resistant sandstone units probably account for 90% of the narrow, twisting canyons of the southern Colorado Plateau, including most of the Escalante Canyons, Zion National Park, and Glen Canyon Recreation Area. Their vertical to overhanging walls form the most constricted and deepest reaches of the Fremont River canyon and its tributaries, as we shall soon see.

30.2 ≈   Almost immediately after entering the Navajo Sandstone the Fremont River spills over a small waterfall in the resistant rock. To the south side of the canyon is what appears to be a large, sand-filled side canyon. This is actually an abandoned meander, part of what used to be the Fremont River channel. Although the abandonment of such meanders occurs naturally, in this particular instance it was caused by the road-builders. Rather than build the road around this sinuous stretch of canyon, they chose to blast through the narrow neck of sandstone. Obviously the river opted for this new, shorter route and straightened its course, deserting its previous route (fig. 8.5). Not surprisingly, the waterfall is the direct result of this event.

Prior to cutting through the meander neck, the river dropped gradually along its sinuous route, giving it a gradient of 35 feet/mile. After abandonment, the drop was the same, but it was abrupt and over a shorter distance (fig. 8.5). The sharp drop is the waterfall.

31.1 ≈   As we continue upstream beneath the steep, unbroken sandstone walls, patches of the wall begin to be pocked by a strange honeycomb pattern

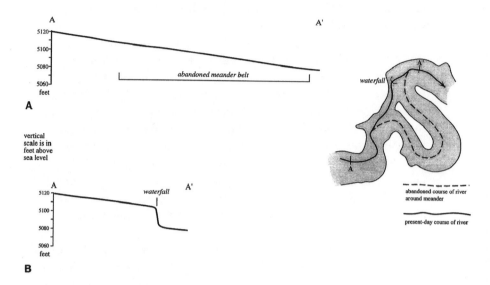

Figure 8.5. Profiles of the Fremont River bed in Capitol Reef showing (*a*) its earlier course as it wound through a meandering canyon, thus producing a low gradient, and (*b*) the present-day course after the Utah Department of Highways cut through the thin neck of the meander, straightening its course and increasing the stream gradient. The drop of the present-day waterfall represents the amount of gradual elevation decrease that the meander previously provided over a greater distance. The modern streambed and the abandoned meander path are indicated on the map.

(photo 9). This is the result of differential weathering of the rock. After the loose Jurassic dune sands were buried by succeeding sediments, chemical-rich fluids passed slowly through the pore spaces between the grains. In the Navajo calcium carbonate ($CaCO_3$) was precipitated into the spaces from these fluids, cementing it into sandstone. This process of turning loose sediment into rock is called lithification.

Some parts within the vast sand pile were better suited for the cement precipitation than others. This may have been due to local chemical conditions in the sediment, or because fluids passing through at that particular time and place were oversaturated with $CaCO_3$, or because local areas with higher porosity promoted the passage of larger quantities of $CaCO_3$-rich fluids. It may have been a combination of these factors or something else. In any case, the quality of cementation varies through the rock.

While the water giveth, it also taketh away. Once cemented rock is exposed to surface waters, the cement begins to dissolve. Acids, however weak, will dissolve calcium carbonate—and all natural surface waters are at least mildly acidic. Additionally, poorly cemented zones in the rock will

Photo 9. Strange erosion pockets in the Navajo Sandstone along the Fremont River in Capitol Reef National Park.

dissolve more rapidly than well-cemented areas, ultimately producing the curious honeycomb pattern seen in the Navajo today. In some places the poorly cemented areas occur along the planes and curves of crossbeds, accentuating their intricate geometries with recesses and ridges.

The vertical black streaks that appear as if black paint had spilled from the cliff tops are thin coatings of desert varnish. These tapestrylike ribbons are the result of manganese- and iron-rich water that drips down the walls during occasional wet periods. In other parts of the canyon the varnish is more widespread and uniform. These areas are believed to develop over longer periods, as miniscule amounts of manganese and iron are slowly leached from the rock to precipitate a molecule at a time on the sandstone face. This process probably takes place over several hundred years, rather than thousands or millions. Evidence for such a time frame lies in the petroglyphs found in the canyon, designs that people of the ancient Fremont civilization pecked into the millimeter-thick varnish layers, to expose the fresh rock beneath. The petroglyphs themselves are partially coated by subsequent varnish. The Fremont people abandoned the area 800 to 900 years ago; thus the "new" varnish that coats the petroglyphs has been accumulating for less than 1,000 years.

32.3 ≈   As we proceed west, the cliffs of the south canyon wall step back momentarily to reveal the mouth of Grand Wash, a steep-walled slot cut into the thick sandstone. The flat, sandy floor of Grand Wash makes a beautiful 2- to 3-mile hike. Above the enclosing canyon walls the golden domes and turrets of Navajo Sandstone that crown the Waterpocket Fold can be seen from the Grand Wash trailhead. The name "Capitol Reef" originates with these domes, which reminded the men of the Powell Survey of the Capitol dome far to the east, in Washington, D.C. These giant beehives of sculptured sandstone dominate the skyline from almost any vantage point in the park.

33.3 ≈   A short distance upstream, a recent large rockfall is evident on the south wall of the Fremont canyon. Fresh, peach-colored rock on the wall is fringed by a pile of jagged sandstone blocks. Large patched areas in the road indicate the original landing spot for some of the large boulders. Rockfalls and slides are part of an ongoing geologic process called *mass wasting*. This particular site provides a firsthand glimpse into one of the processes important to canyon evolution. While water may scour the river channel, slowly cutting deeper, catastrophic rockfalls are continually widening the canyon. This action, in contrast to the monotonous grain by grain removal of sand by rivers, is large-scale, instantaneous, and gravity-driven. The products of mass wasting can be seen everywhere in the blocks of sandstone, large and small, that litter the slopes and floor of the canyon.

34.8 ≈   Proceeding upcanyon, we cross the Fremont River again. After crossing the bridge we see a parking area on the north side of the canyon; this is the trailhead to Hickman Natural Bridge and the Rim Overlook. Stratigraphically we have passed downsection from the Navajo Sandstone into the Jurassic Kayenta Formation. The distinction may not be obvious at first, but the massive, unbroken Navajo cliffs give way to the ledgy lenses of Kayenta sandstone. Although the Kayenta is a cliff-former, it contains thin breaks of siltstone and shale within and around the thicker sandstone lenses.

Jurassic Deserts and Rivers ≈

The Kayenta, which is sandwiched between the two major eolian sand-dune deposits of the Colorado Plateau, also was deposited under arid conditions. Kayenta sands, however, mark the passage of large, west-flowing rivers that traversed an otherwise dry, wind-blasted floodplain. During this time large sand dunes maintained a presence to the north, occasionally encroaching on the vast floodplain. The water and sediment in the

Kayenta rivers originated in the Ancestral Rocky Mountains to the east, a mountainous region that occupied western Colorado and New Mexico and has long since been planed flat.

Like the overlying Navajo, the Kayenta is conspicuously crossbedded, but on a much smaller scale. Instead of 30-foot-thick sets, crossbed sets of the Kayenta rarely exceed 3 feet, a reflection of its fluvial origin. Thin, finer-grained mudstone beds appear to cut indiscriminately across the thicker, more resistant sandstone beds. Mudstones were deposited during low-flow periods in the rivers, when currents were too weak to move the coarser sand that dominates the formation.

Violent floods periodically pounded across the Kayenta floodplain, transforming the tranquil setting into a roiling mass of sediment-laden water. During these floods old channels were filled with sediment and abandoned, while new ones were cut into the floodplain muds. New channels formed as chunks of cohesive mud were ripped from their moorings on the plain. Commonly these clasts of partially consolidated mudstone became part of the sediment in the bottom of the newly carved channel. Such large pebble- or cobble-sized pieces of contemporaneous sediment are called *rip-up clasts*, a testimony to their violent genesis.

Rip-up clasts are nicely displayed in the Kayenta walls about 100 feet up the Hickman Bridge trail. Here, at eye level, large irregular-shaped clasts of laminated siltstone are strewn across the base of a sandstone-filled river channel, relics of a 160-million-year-old flood.

If we continue up the trail, climbing higher into the Kayenta, we encounter a bench veneered by the rounded black boulders, a reminder of an ancient Fremont River. The short (1-mile) trail to Hickman Natural Bridge winds through the upper Kayenta sandstones, which have eroded into a convenient bench. The bridge spans one of the many fins that have been carved into the Kayenta.

These pervasive fins originated as a regional joint system that developed during flexure of the Waterpocket Fold. As these northeast-trending joints became exposed to the rigors of weathering at the surface, they formed weaknesses to be exploited by running water. Runoff was focused into these minute openings, and they were deepened and widened to form the parallel fins seen today. From high on the Kayenta bench, looking south across the Fremont canyon, the joint system is readily apparent.

Eventually, parts of these fins were undermined by running water or by the expansion of rock that was finally released from pressure after millions of years of deep burial, a process called *exfoliation*. As a narrow spot or weakness in the fin wore through, an arch was formed. This is how

Hickman Bridge formed, as did arches and natural bridges throughout the canyon country of southern Utah, including Arches National Park to the east. Joint systems are one of the more significant factors in landscape evolution on the Colorado Plateau, often dictating the orientation and location of canyons.

As we look south from the canyon rim, the gentle eastward tilt of the strata gives the uneasy illusion of standing on a listing ship. At the base of the cliffs across the canyon purple slopes of the Chinle Formation peek from beneath the massive orange walls of Wingate Sandstone. The Wingate cliffs are rimmed by the Kayenta, easily recognized by its ledges and benches that extend far back to the bald white domes and fins of Navajo Sandstone.

35.2 ≈　Back on the canyon floor, as we return to Highway 24 to head west, we leave the Kayenta benches behind and become immersed in the ancient eolian sand piles of the Wingate. The Wingate represents another regional sand sea that grew as the Kayenta rivers were overcome by windblown sand. Like the Navajo, the Wingate is thick and cleaves into cliffs that are slashed by vertical cracks and great, arcing fractures. Giant sandstone slabs and towers lean precariously against the stained and rippled walls, the remnants of once-continuous fins that were generated by jointing and exfoliation. Large sandstone blocks tumbled along the base of the cliffs remind us of the ephemeral nature of steep rock, at least on a geologic time scale.

The beautiful orange-red hues of the Wingate come from the presence of small amounts of iron oxide distributed through the cement that holds the sand grains together. This iron oxide is known to geologists as the mineral hematite. The more hematite in the rock, the deeper the colors. Hematite is the same iron oxide that forms as rust when your monkey wrench is left in the rain.

Along the walls of this deep, shaded section of canyon, veiled by trees, are petroglyphs, the etchings of the long-vanished Fremont culture. These designs—outlines of bighorn sheep and mysterious abstract forms—were pecked into the hard coating of desert varnish with harder rocks, exposing the light-colored sandstone on an iridescent blue-black background.

Anthropologists have long speculated on the fate of the Fremont and their close relatives to the south, the Anasazi. The Hopi people of northern Arizona may hold many of the clues to this mystery. According to Hopi elders, most of the designs seen in petroglyphs on the Colorado Plateau are Hopi clan symbols, marking the passage of these people and documenting certain aspects of the their history. Furthermore, the Hopi claim the Fremont and the Anasazi as their ancestors. A growing body of archeological evidence supports these claims, although some anthropologists remain skeptical. This could mean that the Fremont and Anasazi people who "disappeared" from the region about 800 years ago were simply assimilated into the Hopi tribe.

Also in the shadow of these sheer walls lie the orchards and the remaining buildings of Fruita, an early Mormon settlement. The first homestead in the area was pioneered by Neils Johnson in 1880. By 1900 ten families had settled along the Fremont and established orchards on its floodplain. The families prospered in the mild climate with an inexhaustible supply of irrigation water. In 1937 Capitol Reef National Monument was established, and the homesteads were gradually bought out by the U.S. government. The families slowly filtered out of Fruita, most of them relocating to other towns in southern Utah.

35.6 ≈   As we stop at the "Petroglyphs" turnout (a parking area and trailhead) and look south across the canyon, the colorful Chinle shales begin to emerge from beneath the Wingate, their vivid shades of purple, pink, and green showing through a screen of Wingate detritus. A little farther up the canyon the barren Chinle slopes expand into a splendid counterpart to the orange Wingate walls that loom overhead.

The Chinle was deposited under much different conditions than the arid climate that prevailed over most of the Triassic and Jurassic time. Shales of the Chinle represent deposition in lake and floodplain environments under seasonally wet monsoonal conditions. More resistant sandstone layers mark the passage of large, northwest-flowing rivers that drained a vast area of the continent. These drainage systems had headwaters as far away as modern-day Texas, Oklahoma, and northern Mexico. Chinle rivers met with the Triassic sea in present-day northwest Nevada.

36.0 ≈   After we pass the old Fruita schoolhouse the steep Chinle slopes become increasingly prominent, with their deep hues unfolding into a spectacular palette of colors. At the top of the slope the razor-sharp contact with the Wingate is clearly displayed.

36.2 ≈   As the full thickness of the Chinle is exposed, the brick-red strata of the underlying Moenkopi Formation appear at road level. When we encounter this basal Triassic unit, the canyon opens to reveal a wide floodplain occupied by a patchwork of orchards and lush, grass-covered fields. The orchards are a reflection of earlier days, when Fruita was home to hard-working Mormon families. Today the Park Service maintains the trees, and deer casually graze on the tall grass. The crimson Moenkopi layers dominate the surrounding slopes as we approach the visitor center and Capitol Reef Park headquarters. Like the overlying Chinle, the Moenkopi degrades into steep slopes, although its numerous thin sandstone beds create a rougher steplike profile.

Photo 10. Rippled bedding surface recording the passage of tidal currents in the Triassic Moenkopi Formation, Capitol Reef National Park.

36.4 ≈   The red, horizontal stripes in the Moenkopi are reminiscent of the younger Summerville Formation, whose red strata were seen near Hanksville and the east park entrance. They should be! Both consist of thin interbeds of red sandstone, siltstone, and shale deposited on an arid tidal flat. The lateral continuity of the beds attests to a broad, featureless shoreline, at times inundated by a shallow sea, at other times exposed to the dry air. Uncountable passages of tidal currents are recorded in the abundant rippled sandstone surfaces (photos 10 and 11). Mud cracks and thin gypsum beds mark the periodic recession of water and desiccation of the flat, muddy expanse under the intense Triassic sun. The ripple-marked sandstone slabs used for signs around the park and for the walls of the visitor center come from the ancient tidal sands of the Moenkopi. The formation dominates the vista from the visitor center, with its maroon and orange layers contributing to the spectacular landscape in all directions.

    The view to the south from the visitor center is truncated by a flat, tree-topped mesa. The flat surface contrasts with the surrounding strata that are tipped gently to the east by the upwarp of the Waterpocket Fold. The mesa is built on the tilted Moenkopi, but it has been planed flat by the early Fremont River, which now lies out of view on the south side of

Photo 11. Cross section of ripples in the Moenkopi Formation showing internal cross-laminae, a smaller-scale version of crossbeds.

the mesa. These buckled older rocks are blanketed by a thin veneer of black basalt boulders, deposited as floodwaters of the ancient Fremont gushed from between the confining walls of its narrow upstream canyon. As we saw earlier in terrace deposits in narrower parts of the canyon downstream, the channel floor of the Fremont earlier was at a higher elevation. Subsequent eons of erosion have left this earlier detritus stranded far above the modern river level.

A Side Excursion to the South ≈

The scenic road that extends south from the visitor center provides a magnificent side excursion to the head of Grand Wash, whose mouth we passed earlier in the Fremont canyon. Initially the road winds through dense stands of fruit trees and open grassy meadows that were cultivated by the Mormon pioneers. As we cross the Fremont River, the source of water for this lush greenery becomes apparent. Past the campgrounds the valley opens, and the deep canyon from which the Fremont emits can be seen across the fields to the west. The walls of this canyon are carved from

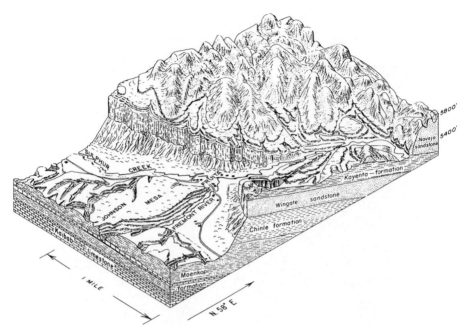

Figure 8.6. A block diagram of Capitol Reef showing the topography and geology where the Fremont River emerges from Miners Mountain to enter the canyon as it cuts through the Waterpocket Fold, following the route of State Highway 24. From Smith and others 1963.

the oldest rocks exposed in the region, the Permian White Rim Sandstone and Kaibab Formation. Fossiliferous Kaibab limestones are the youngest rocks in the much older strata of the Grand Canyon and form the north and south rims of that great chasm to the south. From here the road follows the convenient strike valley cut into the yielding Moenkopi strata. To the west the Moenkopi arches upward to form Miners Mountain, an extensive flat-topped bench. To the east is one of the more striking views in southern Utah's canyon country, the great western rampart of the Waterpocket Fold (fig. 8.6). At road level lie the red strata of the Moenkopi Formation; above that, the steep, multicolored Chinle slopes. This colorful Triassic sequence culminates with the towering walls of Wingate Sandstone. This grand palisade extends northwest for 20 miles, where it is concealed beneath the younger lava flows that cap the Fish Lake Plateau, and another 60 miles southeast to the graveyard of Glen Canyon.

Upon reaching Grand Wash, we may turn east to continue half a mile on a gravel road to the trailhead. An alternative is to continue southward along the strike valley to Capitol Gorge, another major canyon that bisects the Waterpocket Fold. Past Capitol Gorge the road turns southwest onto a four-wheel-drive track that climbs onto Miners Mountain. This track eventually connects with State Highway 12, the north-south route that links Torrey to the north with the town of Boulder to the south (fig. 8.1). Without four-wheel-drive the safest option is to return to the visitor center and continue west on the safe and scenic Highway 24 to Torrey.

9

Capitol Reef National Park Visitor Center
to Torrey via State Highway 24

Departing the visitor center for the town of Torrey, we once again head
west on Highway 24. From the intersection the monolithic form of the
Castle fills our view to the north. Its summit bristles with fingers of
jointed Wingate Sandstone, underlain by giant green, purple, and maroon
Chinle steps. Access to these steps is guarded by fluted columns of laminated Moenkopi strata, their aprons extending out to the roadside.

When we turn west, the Moenkopi layers rise to the southwest, eventually to meet at the monoclinal culmination of the Waterpocket Fold. For
the next three miles the view to the south is one of deeply incised
Moenkopi canyons, testimony to the incessant trickle of water off the fold
crest. Red sandstone slabs scattered along the slopes and ledges display
elaborate ripple marks, the product of a long-disappeared Triassic shoreline.

As we drive west, the thick white sandstone that is the Shinarump
Member of the Chinle makes an abrupt appearance in the lower cliffs to
the northwest, where it caps the seemingly unstable columns of the
Moenkopi. When we trace it along the cliff band, the Shinarump thickens
to the west and thins rapidly to the east until it disappears. Such thinning
is called a *stratigraphic pinchout*. The coarse sand and pebbles of the
Shinarump are the product of northwest-flowing rivers. Its rapid thickness
changes are due to deposition in erosional hollows carved earlier into underlying strata. As rivers continued to course through this incised drainage
network, the hollows gradually filled with sand and gravel until the topography was planed smooth, producing a regional, low-relief plain.
Overlying green shale slopes are the product of huge volumes of airborne
volcanic ash that wafted into the basin from an extensive chain of active
volcanoes lying in parts of present-day Arizona, California, and western
Nevada. The thin brown cliffs that interrupt these floodplain deposits represent relatively placid rivers that meandered across the ash-choked plain.

As we drive to the top of the short hill that borders this cliff band,

Chimney Rock comes into view. This isolated pinnacle is a detached part of the cliff that we have been observing. Chimney Rock is preserved by a precarious cap of Shinarump sandstone that shelters the soft, underlying Moenkopi layers from the gnawing action of dripping rainwater. Immediately ahead lies the Chimney Rock trailhead, where a short trail leads to the top of the Shinarump-capped mesa.

Part of the hiking trail follows the trace of a large fault, a place where two parts of the earth's crust have moved vertically past each other. The fault, which can be clearly seen after walking a short distance up the trail, is located in a saddle to the east through which the trail climbs. Looking eastward into the saddle, we see the brick-red Moenkopi strata to the south juxtaposed with the green Chinle shales to the north. The trail ascends the disrupted and shattered rock of the fault zone. By knowing the relative ages of the rock units involved in the fault, we can clearly determine that the southern, Moenkopi side moved upward relative to the northern, Chinle side. This northwest-trending fault line parallels the road for the remainder of our drive in Capitol Reef, although it is never so clearly displayed as in this saddle.

As we drive west toward the park boundary, the road follows the trace of the fault for about a mile. Along this segment the fault places upper Moenkopi strata against lower Moenkopi—red rock against red rock, making it difficult to trace. Fortunately there are some telltale features along the way. For instance, when the road curves after tracking the obscured fault for a mile the Moenkopi on the south side of the road is clearly tilted in directions that are discordant with the regional dip of the monocline. Such disruption is characteristic of fault zones, where adjacent strata are dragged along the fault surface as movement takes place.

Almost immediately white Shinarump sandstone borders the north side of the road, to the south is the Moenkopi, and somewhere between lies the fault. As we approach the western park boundary the fault once again becomes evident. Smeared against the white Shinarump to the north are red mounds of shattered Moenkopi, heaved upward into a fault contact with the white sandstone. Passing across the park boundary, the road curves south into a sea of Moenkopi. Here we lose the fault as it maintains its northwestward course for several miles before disappearing beneath the young volcanic rocks of the Thousand Lakes Mountain.

Rising through the endless Moenkopi mudstones, we can see the formation beginning to change character. The shaly mounds give way to thicker, continuous sandstone cliffs as we enter the Torrey Member of the Moenkopi. Although not up to the standards of the great eolian cliffs that dominate the surrounding landscape, these sandstone beds are by far thicker than any yet encountered in the Moenkopi. The Torrey, named for the nearby community, represents a large delta that filled this part of the shallow Triassic shoreline. The delta expanded seaward as large, west-flowing rivers fed it fine sand from the diminishing Ancestral Rockies in western Colorado.

As we look ahead in the distance, to the south lies the dark forested bulk of the Aquarius Plateau, also known as Boulder Mountain. This high, volcanic-capped plateau rises to more than 11,000 feet and is the loftiest feature of the High Plateau province of western Utah. To the north sits its sister, Thousand Lakes Mountain. This pair of isolated plateaus at one time formed a single continuous feature. Their origin lies in the youngest rocks in the region, the resistant volcanic rocks that top the plateaus. This durable caprock formed as relatively fluid lava gushed from fissures in the crust, blanketing everything in its path and filling in all preexisting topography. The division of this once-continuous plateau was caused by the slow dribble of water across the flat-topped region, eventually carving the drainage network for the Fremont River, which now bisects the two highlands. After the river labored through the hard basaltic caprock, erosion of the underlying succession of the soft Tertiary Claron Formation was a simple affair, bringing the river close to its present level, where it now winds through the colorful Mesozoic strata that encircle the towns of Torrey and Teasdale.

10

Torrey to Boulder via State Highway 12

This section of our trip begins on the outskirts of Torrey, at the intersection of State Highway 24 and State Highway 12, which we will take south to the town of Boulder (fig. 10.1). Highway 12 traverses the east flank of the Aquarius Plateau and reaches elevations of more than 9,000 feet, offering a welcome respite from midsummer heat (fig. 10.2). Most of the route passes through glades of subalpine aspen and pine forests. Steep, frothy brooks flow eastward off these slopes, eventually to meet with the Fremont River in Capitol Reef and beyond. The deep shadows and abundant greenery are a dramatic change from the bare red and gray rocks of the lower elevations.

As we head south, we will be treated to spectacular views to the east, looking down on the Waterpocket Fold with its precipitous red cliffs and white sandstone domes appearing momentarily where the trees part. Farther back are the sandstone mesas and gray shale slopes of the younger Cretaceous strata. These lap onto the flanks of the Henry Mountains, their great laccolithic bulwarks looking down on the entire Mesozoic succession.

mile

0.0 ≈   As we begin our southward journey to Boulder, we remain in the red deltaic deposits of the Moenkopi Formation. Ahead to the south is the Cocks Comb, a resistant spine composed mostly of Navajo Sandstone. Through the forested lower slopes of Boulder Mountain we can discern the now-familiar white cliffs of Navajo Sandstone as well. The hummocky upper slopes of the plateau are the product of landslides and other mass-wasting processes. These mass movements are initiated in the unstable Tertiary sedimentary rocks that underlie the thin mask of volcanic boulders and loose soil, known to geologists as *colluvium*. This layer of detritus comes mostly from the volcanic caprock of the plateau, in some cases bull-dozed off the plateau summit by glaciers that inched their way from the top over the last 100,000 years (fig. 10.2).

1.1 ≈   After a short distance we encounter a roadcut on the east side of the road, providing an illuminating view into the Torrey delta. This roadcut

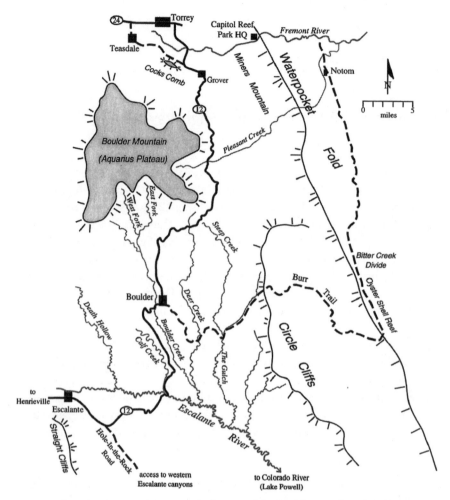

Figure 10.1. Map of the second segment of the road log emphasizing the physiographic features such as highlands and drainages as well as the various towns and roads. The route follows Utah Highway 12 for its entirety. The first part, from Torrey to Boulder, traverses the east slopes of Boulder Mountain. The second part, from Boulder to Escalante, crosses the numerous deep tributaries of the Escalante River that have cut into the Jurassic Navajo Sandstone. The road crosses the Escalante River at its confluence with Calf Creek.

and the thick sandstone body that overlies it demonstrate one phase in the evolution of the deltaic system. At road level irregularly spaced, thin sandstone and siltstone beds suggest an episodic inpouring of sediment, introduced from rivers to the east that flushed plumes of fine sand and silt into the shallow sea. During this time the locus of deposition was located on the main lobe and situated farther east, adjacent to the river's mouth. The

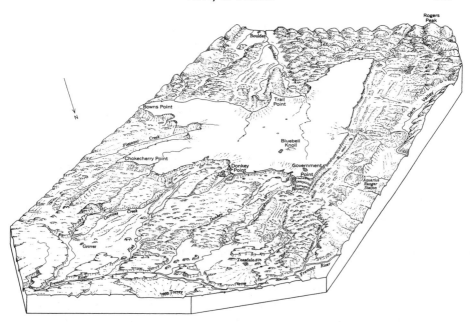

Figure 10.2. Block diagram of the Aquarius Plateau/Boulder Mountain showing the geography, the topography, and the communities of Torrey, Teasdale, Grover, and Boulder. The approximate length of the sides and front of the diagram is 13 miles. The diagram is schematic and not to any exact scale. The view shown is looking east-southeast from some point above Thousand Lake Mountains. From Flint and Denny 1958.

thin-bedded strata before us were deposited at the far, seaward edge of the delta, so that sand only sporadically made it this far. Because the sea was shallow, the area adjacent to the river's outlet filled with sand rather quickly, and no more space was available to accommodate it in this area. At this point the locus shifted seaward, and the delta prograded or built outward. The older part of the delta, which had built up almost to sea level, became inactive, and the river cut across its top, bypassing it to dump its load deeper into the sea. It was this seaward advancement that brought the Torrey delta to where we stand: the advancing lobe is preserved as the thick sandstone cliff that caps the hill above the roadcut.

2.7 ≈   Continuing south, we again intersect the Fremont River, which constitutes the main drainage for this fraction of the canyon country. The headwaters of the Fremont lie on the west flanks of the Fish Lake Plateau; the north and east slopes of the neighboring Aquarius Plateau are also important contributors to its uninterrupted flow. Here the Fremont threads its way through the Moenkopi hills, but quickly disappears to the east on a steep downhill run, where it has slashed deeply into the crest of the great monocline. Along this path it lays bare the oldest rocks seen yet, the

Permian White Rim Sandstone and the Kaibab Formation. After its head-
long plunge through this narrow defile, it emerges near the Capitol Reef
campgrounds, again meeting the red hills of the Moenkopi.

3.7 ≈   As we climb from the Fremont River through the Moenkopi, the spine
of the Cocks Comb appears, jutting skyward to the southwest. The craggy
fin is composed of steeply tilted, fractured Navajo Sandstone. In its northern shadow beneath the rubble lie the shattered remnants of the Kayenta,
Wingate, and Chinle Formations (fig. 10.3), also sharply tilted to the
south. Across the valley to the north sits the relatively flat-lying Moenkopi.
This abrupt transformation is a reflection of the Teasdale fault zone, which
occupies the intervening valley. This zone of folding and faulting forms an
extensive, 2-mile-wide belt of disruption. From the Cocks Comb the zone
extends 15 miles to the southeast. In this direction the fault cuts through
all the strata to be exposed at the surface. To the northwest the fault
pushes another 10 miles, passing through the community of Teasdale,
from which the zone gets its name. Throughout this segment, however,
no faults break the surface. Instead the zone is expressed as sharp folds
that, in the Cocks Comb area, shove the Moenkopi on the north upward,
to the level of the younger Glen Canyon Group rocks. This upward push
on the northeast side of the Teasdale zone dragged the edge of the southwest side up with it, tipping the rocks on the Cocks Comb southward.

4.6 ≈   Another mile to the south the turnoff to Teasdale appears on the west.
This road adopts the trace of the Teasdale fault zone, following it to the
small community. From there the road diverges northward back to
Highway 24. The disrupted fold and fault zone continues northwest and
dies out in several miles.

For the next few miles we wind through the volcanic detritus derived
from the top of Boulder Mountain and across the narrow valley that constitutes the hamlet of Grover. The hummocky hills around Grover consist
of glacial till, a chaotic mixture of clay- to boulder-sized fragments. The
debris was bulldozed down the flanks of the Aquarius Plateau by glaciers
that spilled from its top about 125,000 years ago. This lobe-shaped
deposit originates at the cliffs that cap the plateau and spreads as far
downslope as the Fremont River. North of Grover, however, the rubble is
dissected by erosion, and only isolated remnants remain.

The angular black boulders that dot the hillsides are the same basalt as
the spherical "cannonballs" seen on the Fremont River terraces in Capitol
Reef. These differences illustrate one of the basic principles of geology—
the greater distance a particle has traveled in the sedimentary cycle, the
rounder and smaller it becomes. The degree of rounding is a reflection of
the amount of abrasion a particle experiences on its downhill journey.

southwest   northeast

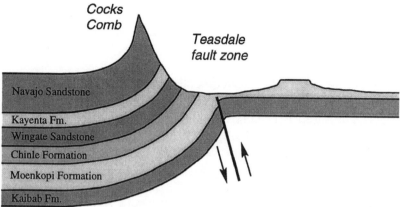

Figure 10.3. Cross section of the Cocks Comb area showing the relationship of the Teasdale fault zone, the Cocks Comb, and the various strata involved in the deformed area. The diagram is purely schematic, and no horizontal or vertical scales are implied.

Along this segment the highway and the small valley in which Grover resides follow the Teasdale fault zone. The fractured and crushed rock generated in these disrupted zones makes them easy prey to erosion, producing topographically low areas amidst otherwise resistant rock. Once again the enduring team of gravity and water has excavated a convenient pathway.

7.1 ≈   After passing through Grover we begin our climb through the pygmy forest of piñon and juniper into the stately pines and fir. On the east side of the road is a cut that exposes bleached and tilted Moenkopi that has been caught in the convulsive displacement of the fault zone. Here the beds have been tilted 30° to the southwest.

7.6 ≈   As we continue up the hill, bluffs of white Navajo Sandstone show through the increasingly dense forest. We have veered from the trace of the fault zone as it bears southeast through the valley below, bounded by the red Moenkopi to the northeast and the white scarp of Navajo Sandstone to the southwest.

8.3 ≈   The irregular creases in the Navajo give it a curious appearance resembling an elephant hide. This peculiar texture is the consequence of joints and the incessant trickle of water along them. Tree-covered slopes above the Navajo are mostly the Jurassic Carmel Formation.

9.8 ≈   Curving up the hill, we can see to the east above the road the red and white gypsum-rich strata of the Carmel. Thin-bedded orange-red siltstone was deposited by oscillating tidal currents that transferred fine sediment along the extensive shallow shoreline. Thicker gray and white siltstone and gypsum record fluctuating conditions on the tidal flat. Gypsum beds attest to brines rich in calcium sulfate, conditions that would be deadly to most organisms. Gypsum probably formed in extensive shallow evaporative pans, essentially large puddles of seawater abandoned by the sea as it receded for a geologic moment.

Below the Carmel, blanketed by sand, rock, and pine needles, the sweeping curves of the Navajo Sandstone are revealed. The crests of these bygone dunes were blunted by the mowing action of erosion prior to Carmel deposition. This is clearly shown by the razor-sharp planar contact between the disparate rock types.

10.9 ≈   A closer examination of the Carmel can be had a short distance up the road, where white laminated sandstone and siltstone are bared in a road-cut. The thin, even laminae suggest the uneventful accumulation of sediment under quiet, shallow-water conditions.

13.8 ≈   After driving several miles with our view denied by the thick stands of aspen and fir, we turn into the Larb Hollow overlook, where a stunning vista unfolds to the east. Below lie several parklike areas nestled into the Navajo Sandstone. We look down on Pleasant, Tantalus, and Vergy Flats, green openings severed from each other by the deep gash of a canyon or by serrated sandstone fins. Behind this the Moenkopi forms the broad backbone of Miners Mountain. The Moenkopi plunges eastward off the serrated crest of Waterpocket Fold to disappear into the subsurface. On the skyline the ubiquitous Henry Mountains tower above it all, their snow-capped summits dominating the panorama. The scene has not changed since Dutton, in his 1880 monograph *Geology of the High Plateaus*, so accurately described it from a vantage point at the rim of the Aquarius Plateau:

> To the northeastward the radius of vision reaches out perhaps a hundred miles, where everything gradually fades into dreamland, where the air boils like a pot, and objects are just what our fancy chooses to make them. Perhaps the most striking part of the picture is in the middle ground, where the great Water Pocket fold turns up the truncated beds of the Trias and Jura, whose edges face us from a great quadrant of which we occupy the center. Where the strata are cut off in this way upon the slope of the monocline they do not present to the front a common cliff and talus with a straight crest-line, but a row of cusps like a battery of shark's teeth on a large scale.

For several more miles the road traverses the dense forest, which grows on a combination of gravity-driven deposits of both glacial and landslide origin. Glacial till consists of an unsorted chaotic assemblage of clay- to boulder-sized sediment pushed down the slope by glaciers. Mostly, however, it is landslide debris that we cross, the simple product of gravity acting on unstable slopes.

22.3 ≈ After about 10 miles of forested slope we cross a ridge and are able to focus our gaze southward, to new horizons. To the southeast lies the red rim of the Circle Cliffs, mostly Wingate Sandstone that rises from incisions into the crest of the Waterpocket Fold. White slickrock of the Navajo Sandstone marks the Escalante drainage system. Myriad canyons and hollows originate on the south and east flanks of the well-watered Aquarius Plateau, twisting their way to an eventual rendezvous with the main stem of the Escalante River. This great trunk stream flows south to meet the stagnant waters of Lake Powell. Looking in that direction, the eye is drawn to the broad dome of Navajo Mountain, in whose shadow lies Rainbow Bridge, a gargantuan natural bridge of streaked red sandstone. Extending northwest from the hulk of Navajo Mountain is the long, striated bluff of the Straight Cliffs, also known as Fiftymile Mountain. Behind this the Kaiparowits Plateau begins. These scenes continue through the thinning trees until the road drops to head due south into the town of Boulder.

30.1 ≈ As we enter the Boulder town limits, it initially appears as if there is no town at all—only scattered ranch houses. Soon, though, houses become more closely spaced and a town materializes.

35.2 ≈ Anasazi State Park is located within the town of Boulder. The park consists of a museum with numerous well-preserved relics of the Anasazi and several excavated dwelling sites. It seems that the Anasazi found this valley as appealing as its modern inhabitants do.

Boulder, with its lush green fields fed by the sparkling waters of Boulder Creek hemmed in by the honey-colored sandstone cliffs and fins, is the most strikingly beautiful settlement in southern Utah. It has been, and remains, a ranching community. Its fantastic setting and its location between Capitol Reef and Grand Staircase–Escalante Monument may change things, however—I hope not. For now, its open pastures remain as pristine as they were 100 years ago.

The first permanent settlement in the valley was established in 1889 when the family of Amasa and Roseanna Lyman established a place in what today is upper Boulder. Their trail into the valley from Grover was fraught with obstacles in the form of heavy boulders and trees that had to be cleared for the passage of their wagons. After settlement the town remained extremely isolated for many years thereafter. Even into the 1930s Boulder received mail only twice a week, brought in from Escalante by mule. The modern, year-round road between Boulder and Escalante that today is State Highway 12 was finished by the Civilian Conservation Corps (CCC) only in 1940 and was not paved until 1971.

This segment of the road log ends at the intersection between Highway 12 and the Burr Trail, a road that traverses the Circle Cliffs and provides access to many of the eastern tributary canyons to the Escalante (fig. 10.1). The road continues eastward across the crest of the Waterpocket Fold in Capitol Reef and down through steep, bumpy switchbacks to meet with the Bullfrog–Notom road, which parallels the Waterpocket Fold to go either north to Highway 24 through the Waterpocket Fold district of Capitol Reef or south to Lake Powell.

11

Boulder to Escalante via State Highway 12

The region that we will traverse on this segment contains some of the most mysterious and enchanting canyons on the face of the earth. As Dutton (1880) gazed over this vast Escalante basin in the 1870s, from the south ramparts of the Aquarius Plateau, a breathtaking view unfolded before him:

> It is a sublime panorama. The heart of the inner Plateau Country is spread before us in a bird's-eye view. It is a maze of cliffs and terraces lined off with stratification, of crumbling buttes, red and white domes, rock platforms gashed with profound cañons, burning plains barren even of sage—all glowing with bright color and flooded with blazing sunlight. Everything visible tells of ruin and decay. It is the extreme of desolation, the blankest solitude, a superlative desert.

mile 0.0 ≈   The trip from Boulder to the town of Escalante continues west on State Highway 12. This stretch of road may be one of the most spectacular in the United States and, for the most part, lies within the Grand Staircase–Escalante Monument.

As we leave the intersection of the Burr Trail and Highway 12, the view is to the west, toward the elaborately crossbedded and jointed Navajo Sandstone that rims the west side of Boulder valley. The steplike, tree-covered red slopes above are in the Carmel Formation.

1.9 ≈   Leaving Boulder, we grind upward through the Navajo roadcuts to the top of the hill, where the Carmel is well exposed. Here the formation consists of alternations of crossbedded eolian sandstone and thinly bedded red and white siltstone that record incursions of the Sundance seaway. These contrasting systems were in competition as the shallow sea extended its thin fingers southward into the coastal dune field.

3.0 ≈   Soon after reaching the mesa top the Hell's Backbone road branches sharply to the right. This adventurous dirt backroad is the original route to the town of Escalante and skirts the headwaters of a large part of the Escalante River drainage. This was the first road connection between

Escalante and Boulder and, like the others, was built by the CCC. It was not a year-round route, however, as it climbs to the dizzying head of Death Hollow at an elevation of more than 9,000 feet. The CCC began work on the modern route of Highway 12 soon after completing this circuitous road.

The next 3 or 4 miles take us through the piñon-juniper forest at the top of the mesa, called New Home Bench. Black volcanic boulders blanket the mesa top, just as they cover all the surrounding mesas. These resistant rocks derive from the Aquarius Plateau and represent a time several million years ago when the flanks of the plateau were young and undissected. During this time an apron of alluvium fringed the lofty plateau—today only these remnants cap the last vestiges of this once unbroken surface.

7.0 ≈   As we pass onto the narrow isthmus of New Home Bench, the walls plunge sickeningly on either side. The view is breathtaking; the floor of Dry Hollow, a tributary to Boulder Creek, lies 500 feet below to the east. On the west side and 600 feet below sits Calf Creek, a delightful tributary to the Escalante that we will soon drop into. Hopefully not too soon. Across the vast maze of Navajo Sandstone to the west can be seen Powell Point at 10,188 feet, the southern promontory of the Table Cliff Plateau. In reality the Table Cliffs are a southern finger of the Aquarius Plateau. To the southwest the Straight Cliffs again are in view. This great wall of Cretaceous strata forms the eastern boundary of the lower Escalante drainage net.

8.0 ≈   As the road drops from the mesa top, we again descend through Navajo roadcuts. Sweeping large-scale crossbeds are clearly displayed for those who are able to remove their eyes from the road.

10.7 ≈ On our descent to Calf Creek, red shale and siltstone of the Kayenta Formation appear beneath the Navajo. The contact between the two is well exposed in a roadcut on the south side of the road. The fluvial Kayenta is sandwiched between the eolian Wingate Sandstone below and the Navajo above. The tripartite assemblage forms the Glen Canyon Group. The west-flowing Kayenta rivers originated in the remnant highlands of the Ancestral Rockies in present-day western Colorado. In contrast to its cliff-forming eolian neighbors the Kayenta has a distinctive ledgy appearance, the expression of large nested, lens-shaped bodies of sandstone representative of river channels, separated by thin, less-resistant siltstone and shale layers.

Calf Creek ≈

11.2 ≈ The Calf Creek campground and trailhead appear on the north side of the highway. This scenic stop is run by the Bureau of Land Management (BLM) and consists of developed camp sites and a beautiful hiking trail along Calf Creek. The trail meanders upstream through the sagebrush and cottonwoods along the canyon floor. For the duration of the hike the Kayenta remains at river level. After 2.75 miles the trail ends at an impassable grotto, sliced at its head by a ribbon of water cascading down a moss-covered, water-polished slide of Navajo Sandstone. Surrounding black-streaked sandstone walls are accentuated by clumps of ferns and wildflowers that appear to be squeezed from dripping cracks. This shaded alcove, with its rare luxuriance of vegetation, is permeated by a fine mist that comes as a welcome respite from the surrounding rugged and parched environment. This short, scenic hike provides a rewarding break for anybody passing through the region.

11.7 ≈ Where the highway crosses Calf Creek the complex internal geometry of the Jurassic river channels can be seen at eye level, clearly etched into the canyon walls (photo 12).

Photo 12. Overlapping, stacked sandstone-filled river channels of the Jurassic Kayenta Formation along the Escalante River. The massive cliff-forming sandstone above is the Navajo Sandstone.

Escalante River ≈

12.3 ≈   As we cross the Escalante River, the Kayenta surrounds us, but everywhere the sentinel-like cliffs and domes of Navajo Sandstone overlook the
river. This is the single place that a highway crosses the Escalante River: trails
head both upstream and downstream from here, entering a vast redrock
wilderness in either direction. The Escalante is one of the larger rivers of the
southern Colorado Plateau, yet so tortuous are its canyons and the country
it passes through that both of John Wesley Powell's exploratory forays into
the region failed to recognize it. Its drainage net is vast, covering an area of
1,840 square miles, including the southern Aquarius Plateau, most of the
expansive Circle Cliffs, and the northeast part of the Kaiparowits Plateau.
Probably 90% of the river (and its tributaries) is rimmed by red sandstone of
the Glen Canyon Group. The phenomenal maze of canyons makes the
Escalante region unique on this planet.

13.2 ≈   After crossing the river the road rapidly climbs upward out of the
canyon, passing through the Kayenta and back into the never-ending
Navajo. Near the top of the hill Boynton Overlook provides a convenient
vantage point from which to view the Escalante. Across the canyon are
orange and white domes of the Navajo. Below, at river level, are the more
variable broken red crags of the Kayenta. Looming on the horizon far to
the north is the southern escarpment of the Aquarius Plateau.

For the next 3 miles the road undulates across ridges and swales of
Navajo Sandstone. Upon being immersed in this strange landscape we
may begin to imagine the scenes of the immense sand seas of the Jurassic
Period in this region. These bowls of bare sandstone probably contain the
finest exhibition of large-scale eolian crossbedding in the world. The erosion patterns in some of these hollows produce three-dimensional fins that
reflect the sloping front or *slipface* of the colossal ancient dunes. These are
the surfaces along which the sand grains, hundreds of millions of years
ago, rolled and slid.

16.9 ≈   Following a short climb through switchbacks in the Navajo, the road
flattens across the top of the formation. Thin red shale beds and eolian
sandstone of the Carmel greet us at the top. From the scenic turnout we
can look out over our previous route, as though the entire southern
Colorado Plateau has been laid at our feet. The laccolithic summits of the
Henry Mountains can be seen 40 miles to the east. To the north lies the
flat-topped Aquarius Plateau, the origin of most of the drainages we have
crossed since leaving Boulder. Sixty miles to the southeast, across the
drowned Colorado River, is Navajo Mountain. Most of the rocks exposed
between this overlook and these igneous highlands are Navajo Sandstone,

in some cases capped by thin remnants of the Carmel Formation in which we now find ourselves.

18.0 ≈   After we pass through several roadcuts in the Carmel, in which we will remain until the town of Escalante, a commanding panorama of the loftier areas of Grand Staircase Monument unfolds. The Straight Cliffs, guardian of the Kaiparowits Plateau, form a daunting barrier to the southwest. Beyond that is Canaan Peak, at 9,293 feet the high point of the Kaiparowits. Most of the plateau is composed of a bewildering mix of gray and brown sandstone, shale, and coal. This diverse assemblage of marine, fluvial, and swamp deposits owes its genesis to oscillations of the Cretaceous Western Interior seaway and recurring uplift in the mountainous sediment source in western Utah.

Canaan Peak has an intriguing history. The entire summit of the peak is a landslide block of the Canaan Peak Formation that apparently slid into place as a gigantic gravity-driven mass of rock. Today the high-standing mass is surrounded by older rock at a lower elevation; thus it is unsuitable as a source for such a catastrophic event. The slide must have occurred long ago, probably several millions of years ago, although there is no way to determine the age accurately. It must have happened before regional erosion had reduced the area to its present elevation and locally removed the Canaan Peak Formation, however. An interesting problem indeed!

Directly ahead on the skyline is the Table Cliff Plateau. The southmost extension of this alpine mesa is Powell Point, which reaches an elevation of 10,188 feet. The lower gray slopes are the northern continuation of carbon-rich siltstone and shale of the Kaiparowits Formation. The upper slopes are composed of the Canaan Peak Formation, which straddles the Cretaceous and Tertiary Periods. Overlying this are cliffs of the Tertiary Grand Castle and Pine Hollow Formations. Both consist of fluvial sandstone and conglomerate derived from localized uplifts that formed during the Laramide orogeny. Sediments that make up these formations accumulated in small basins that developed between the uplifts. Capping these units, and forming the top of the plateau, is the upper part of the Claron Formation. The Claron was deposited after tectonic activity of the Laramide and represents relatively quiet-water lake deposition as the small basins slowly filled with sediment.

Hole-in-the-Rock Road ≈

21.7 ≈   The road to Hole-in-the-Rock and access to the numerous western tributaries of the Escalante River turns south. This graded dirt road parallels the Straight Cliffs for their entirety and provides an entrance to some of the most spectacular canyons of the Escalante system. The road ends after 55 rough miles at Hole-in-the-Rock, a narrow slot carved through cliffs of Glen Canyon Group sandstone by Mormon pioneers in 1879. The unlikely route was forged as a crossing of the Colorado River in an effort to establish a wagon trail to the rugged southeast corner of the state. These Mormon families, who spent a hard winter on the Hole-in-the-Rock trail, were sent by the church to establish a settlement. After they forged the route and lowered wagons by rope to the banks of the Colorado River, it was not used again, at least not by wagons. Today the top of the route can be seen at the end of the rough road; the lower part lies beneath the oily waters of Lake Powell.

23.3 ≈   To the north bedding in the Carmel Formation begins to tilt noticeably to the southwest, the first hint of deformation since leaving Waterpocket Fold. Not surprisingly, this structure called the Escalante monocline has a similar heritage. It is another expression of the Laramide orogeny, the end result of intense compression of the earth's crust 65 to 50 million years ago. This event threw the sedimentary layers of southern Utah into a series of north-tending folds, like wrinkles in a gigantic rug. The view of this particular wrinkle improves to the north as we enter Escalante.

Escalante Monocline ≈

25.5 ≈   As we approach Escalante, the Escalante River trailhead turnoff is to the right, just before the graveyard. Immediately to the north is the wide Escalante River valley, a lush meadowland where hay is cultivated. Here the river runs eastward as it cuts through the monocline. Farther to the north lies the valley of Pine Creek, a tributary through which the Devil's Backbone road enters the town. Looking up this valley, we obtain an incredible view of the Escalante monocline (fig. 11.1). On the east side of the valley are the rugged, tawny slopes of Navajo Sandstone, dotted with mammoth ponderosa pine trees. Here the Navajo takes a sharp westward dive into the subsurface on the limb of the great fold, not to resurface until about 65 miles to the west in the vicinity of Zion National Park, where it again dominates the landscape. On the west side of the valley are the overlying Jurassic rocks, which from bottom to top (stratigraphic order) include the Entrada Sandstone and the Morrison Formation, all capped by the Dakota Sandstone. The Dakota is the beginning of the

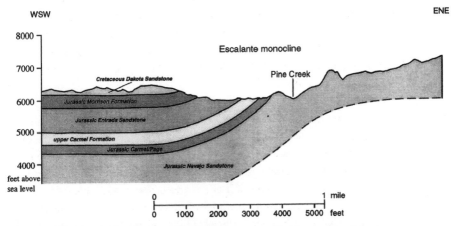

Figure 11.1. Cross section through the Escalante monocline just north of the town of Escalante. After Williams and others 1990.

great succession of Cretaceous rocks that makes up the Kaiparowits Plateau and forms the foundation of the Paunsaugunt Plateau, which we will soon be traversing.

The Town of Escalante ≈

The broad valley that the town of Escalante occupies was called Potato Valley by the first reported party of Anglos to visit the area in 1866. The name derives from the "Indian potatoes," a native plant with sweet, edible roots, that grew on the valley floor. In 1872 A. H. Thompson and F. S. Dellenbaugh, with other members of John Wesley Powell's second Colorado River expedition, entered the valley on their overland journey from Kanab to the mouth of the Dirty Devil River. Upon encountering the unknown and unnamed stream that coursed through this valley, they realized that this was the stream that ultimately provided the impetus for their rugged traverse.

The story began when the great Mormon scout Jacob Hamblin was contracted by Powell to deliver provisions to the mouth of the Dirty Devil River to resupply his expedition. Unfortunately, Hamblin mistook this unknown river for the Dirty Devil; after 50 miles of slogging through narrow winding canyons, often floored by quicksand, he dumped the much-needed supplies at its mouth instead. Thompson and Dellenbaugh were returning to the Dirty Devil to retrieve a boat that they had earlier cached when the supplies failed to be delivered to its mouth. Upon recognizing *this* stream they realized what Hamblin had done, as it had also initially fooled them into thinking they were following the Dirty Devil River. Thompson conveys it best (after climbing to the top of the Escalante monocline for a view):

> On reaching the summit we found we were on the western rim of a basin-like region, 70 miles in length by 50 in breadth and extending from the eastern slope of the Aquarius Plateau on the north to the Colorado River on the south and from the Henry Mountains on the east to our point of observation on the west. A large portion of this area is naked sandstone rock, traversed in all directions by a perfect labyrinth of narrow gorges, sometimes seeming to cross each other but finally uniting in a principal one, whose black line could be traced, cutting its way to the Colorado a few miles above the mouth of the San Juan River. Away to the east and 50 miles distant rose the Henry Mountains, their gray slopes streaked with long lines of white by the snow which yet remained in the gulches near their summits. On our voyage down the Colorado River in 1871 we had determined the mouth of the Dirty Devil to be about 30 miles northeast from these mountains, making it at least 80 miles from our present camp and directly across the network of canyons before us. To proceed farther in the direction we had been pursuing was impossible. No animal without wings could cross the deep gulches in the sandstone basin at our feet. The stream which we had followed and whose course soon became lost in the multitude of chasms before us was not the one we were in search of but an unknown, unnamed river, draining the eastern slope of the Aquarius Plateau and flowing through a deep, narrow canyon to the Colorado River. Believing our party to be the discoverers, we decided to call this stream in honor of Father Escalante, the old Spanish explorer, Escalante River and the country which it drains Escalante basin. (Gregory 1939)

While in the valley Powell's men happened onto a group of Mormons from the settlement of Panguitch who were considering a new settlement. Thompson and Dellenbaugh suggested they call it Escalante, after the newly christened river.

The first families arrived in 1875 and chose the town site in Potato Valley, at the junction of Pine Creek and the Escalante River. It was decided late in 1876, however, that the town site should be moved to the bench just south of the river valley, so the rich floodplain could be utilized for farming. The town stands on this bench today. The town grew rapidly at first and by 1882 had a population of more than 400. Most of the people came from the higher-elevation settlements in the High Plateau region to the west seeking a milder climate.

Today the town may be getting an unwelcome revitalization through the establishment of the Escalante–Grand Staircase National Monument, which now encircles the town. The town natives fear the loss of their old way of life, which dominantly consists of harvesting a dwindling lumber supply from the surrounding high country and cattle ranching on the mostly overgrazed public lands. Only time will tell the future of the town of Escalante.

The Disappearance of Everett Ruess ≈

On November 12, 1934, a young man of twenty-one by the name of Everett Ruess headed south out of the town of Escalante with two burros loaded with supplies. His destination was the maze of slickrock canyons along the lower reaches of the Escalante River and beyond. They moved deliberately toward the Hole-in-the-Rock, exploring the heads of the various canyons that intersected their path. On November 19 he reached the head of Soda Gulch, where he met two sheepherders with whom he camped for two nights. After questioning them at length about the strange and desolate landscape that lay ahead, he continued toward Hole-in-the-Rock. They were the last to see Everett Ruess.

In early March of the following year searchers discovered his burros quietly munching grass on the floor of Davis Gulch, a serpentine, grotto-filled canyon near the terminus of the Escalante. There was no sign of Everett Ruess—he had vanished into the tortuous glens of the lower Escalante. Countless searches since then have turned up little more than the word "NEMO" carved into the sandstone of Davis Gulch, which is attributed to Ruess.

Everett Ruess was a budding young artist who loved nothing more than wandering the arid Colorado Plateau country with his burros. His letters to parents and friends written during these forays documented his intense passion for these wildlands. His blockprints are cherished mementos of this footloose young man who has gained folk hero status among those who love these canyons today. He wrote eloquently of being consumed by this beautiful country, and it appears to have happened. In one of his last letters before departing Escalante he wrote: "As to when I shall visit civilization, it will not be soon I think. . . . I have known too much of the depths of life already, and I would not prefer anything to an anticlimax. That is one reason why I do not wish to return to the cities" (Rusho 1983). While there remains much speculation over his exact fate, such words heighten his legendary status. In any case, anybody who has heard or read of Everett Ruess finds it difficult to ignore him when viewing the cool, shadowed recesses of the lower Escalante canyons.

12

Escalante to the Bryce Canyon Turnoff
via State Highway 12

This leg of our journey begins at the west edge of the town of Escalante and concludes at the turnoff to Bryce Canyon National Park, a driving distance of 45 miles (fig. 12.1). The beginning of our route will traverse the northern part of the Kaiparowits Plateau, skirting the southern margin of the Table Cliff Plateau. State Highway 12 begins in Escalante at an elevation of less than 6,000 feet and rises to about 7,500 feet before dropping again to 6,000 at the Paria River. After crossing the river and passing through the town of Cannonville the route turns north through the town of Tropic. From Tropic the road rises up the east flank of the Paunsaugunt Plateau to top out at an elevation of about 7,700 feet.

We will be leaving the Escalante River drainage and entering that of the Paria River. The Paria is a smaller stream that drains into the Colorado River at Lee's Ferry in northern Arizona, where river trips through the Grand Canyon begin. Like the Escalante River, the Paria cuts the colorful Mesozoic rocks through deep, narrow canyons.

Although this segment of the trip begins in the colorful Jurassic rocks, most of the road passes through Cretaceous strata, drab-colored sandstone and siltstone with lesser shale and coal. Jurassic rocks dive westward into the subsurface on the Escalante monocline, not to resurface until Henrieville, 25 miles to the west, where they are uplifted on the Johns Valley anticline. The final leg up the Paunsaugunt again traverses Cretaceous rocks, but the last uphill grade culminates in the multihued Tertiary rocks that make Bryce Canyon such a spectacle.

The various Cretaceous units encountered on the Kaiparowits Plateau are difficult to tell apart, even with the trained eye of the geologist. Rock types and colors are similar throughout the succession. If our location within the thick sequence is carefully tracked, however, beginning with the basal Dakota Sandstone, recognition becomes straightforward. The road log that follows will help considerably in this task.

As you may recall, recognition of the different Cretaceous formations

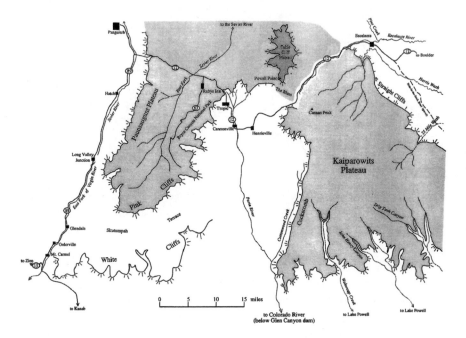

Figure 12.1. Map for the geologic road log showing the route from Escalante to Bryce Canyon National Park and continuing to Mt. Carmel.

near the Henry Mountains is a fairly simple affair since they are defined by obvious changes in rock type. In the Kaiparowits area, however, most of the units are dominated by sandstone and defined by subtle differences such as an increased amount of siltstone or shale. Fortunately these changes are reflected in the canyon walls—thick sequences of sandstone and conglomerate tend to form massive cliffs, while sandstone with minor amounts of shale or coal appears as thin, ledgy cliffs broken by slopes. These outcrop characteristics can be used to recognize the different units and, ultimately, help to decipher their origins.

High sandstone cliffs along Highway 12 are the product of energy-charged, east-flowing rivers that funneled large volumes of sand and gravel to the waiting sea. Ledgy slope-forming sequences were mostly deposited by sluggish, meandering streams. Associated coal formed in vegetation-choked swamps in low-lying coastal areas, similar to the swamps of southern Louisiana today. These environments differ from those active at the same time in the Henry Mountains region, where marine processes dominated the scene.

It is helpful to review the general Late Cretaceous paleogeography for the Colorado Plateau. An extensive north-south–trending mountain chain, called the Sevier orogenic belt, lay to the west along the Nevada-Utah border. Paralleling these mountains to the east was the Western Interior seaway, a linear ocean that extended from the modern Arctic Ocean southward to the Gulf of Mexico. Shorelines shifted substantially during this time, subject to the whims of sea level. At times the shoreline extended far west into central Utah; at other times it receded eastward to the Colorado-Utah border. Sea-level changes and variations in the volume of sediment shed off the Sevier mountain belt were the primary controls on the deposits seen today.

The Henry Mountains area was inundated by sea for much of Late Cretaceous time. When sea level rose, water deepened, resulting in thick deposits of shale. When sea level dropped, or a large pulse of sand and gravel was flushed eastward from the mountains, extensive deltas built seaward. In contrast, the Kaiparowits was situated farther inland from the seaway. Although the sea did reach the area, it rarely was deep enough to produce the shale deposits seen to the east. Instead sea level affected the energy level of rivers passing through the area. If sea level rose, these rivers became sluggish, and swampy wetlands developed. It was the coal deposits of these swamps that made the inclusion of the Kaiparowits Plateau in the Grand Staircase–Escalante National Monument so controversial. When sea level dropped, rivers were rejuvenated and cut through the area on their seaward journey, marking their passage by deposits of sandstone and conglomerate.

mile
0.0 ≈   The Interagency Visitor Center at the western edge of Escalante is the local headquarters for Grand Staircase–Escalante National Monument, the U.S. Forest Service, the Bureau of Land Management, and the National Park Service. Information, maps, and literature for the region are available here.

Looming over the visitor center parking lot are broken sandstone cliffs of the Salt Wash, underlain by the steep varicolored slopes of the Tidwell; both are members of the Jurassic Morrison Formation (fig. 12.2). To the north, across the floodplain of the Escalante River, the white Entrada Sandstone barely shows above the irrigated fields. The red slopes above are the Tidwell with a cap of Salt Wash. The resistant Salt Wash is comprised of fluvial sandstone and conglomerate derived from mountainous areas that lay far to the southwest.

0.3 ≈   In the gulch to the south shallow marine strata of the Tidwell Member are particularly well exposed. The thinly bedded red, green, and white

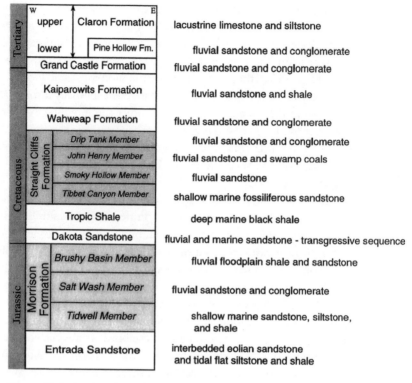

Figure 12.2. Stratigraphic column showing the units encountered between the town of Escalante and the Bryce Canyon area along State Highway 12.

siltstone and sandstone are sheltered beneath a cap of Salt Wash. At road level is the Entrada Sandstone; its large-scale crossbedding is evidence of its eolian origin.

0.8 ≈   The turnoff to the north leads to Escalante Petrified Forest State Park, situated about 0.5 miles up the road next to the Wide Hollow Reservoir. The abundant silicified wood occurs in the Late Jurassic Morrison Formation. Petrified logs indicate a shift in the Jurassic climate from an earlier, eolian-dominated setting to increasingly wetter conditions.

1.3 ≈   Purple and green slopes to the south are floodplain mudstones of the Brushy Basin, the uppermost member of the Morrison Formation. The yellow-brown cliff band above the Brushy Basin is the Late Cretaceous Dakota Sandstone, the basal unit of the thick Cretaceous succession (fig. 12.2). The Brushy Basin is thinner here than to the east in Capitol Reef, where it was last exposed. Such westward thinning is attributed to Early Cretaceous uplift. Based on the depth to which pre-Dakota erosion has occurred in western Utah, the amount of uplift was greater to the west, toward the Sevier orogenic belt, to which it is probably related. Another 25 miles to the west, along the west margin of the Kaiparowits, the entire Morrison Formation is absent, and the Dakota lies on the Entrada Sandstone. The Brushy Basin–Dakota contact here is an unconformity that marks a gap in the record of more than 50 million years.

2.4 ≈   In the roadcut on the south side is a well-exposed stream channel in the Brushy Basin. This lens-shaped body of conglomeratic sandstone displays complex crossbedding typical of river deposits. Pebbles are dominantly red and green chert, a composition that has earned it the name "Christmas tree" conglomerate.

2.8 ≈   The west-plunging, resistant rib of rock on the south is the top of the Dakota Sandstone, the beginning of the immense Cretaceous succession and our introductory reference point. The base of the Dakota, which is covered here, is a slope-forming sequence of coaly shale and carbon-rich sandstone. Black clots of carbonaceous plant fragments are pervasive in these rocks. Sandstone and conglomerate that normally mark the base of the Dakota are missing here. The upper Dakota is composed of fossiliferous sandstone. Fossils are abundant and include oyster beds, *Gryphaea*, and other assorted bivalves, all pointing to a shallow-marine environment.

In most occurrences three distinct parts are recognized in the Dakota. The basal sandstone and conglomerate, which locally are missing, were deposited by vigorous east-flowing rivers, the first hint of the rising Sevier mountain belt to the west. A short distance to the east lay the Western Interior seaway, which was slowly advancing. As sea level climbed, the lower reaches of the rivers were flooded and replaced by muddy coastal swamps. It was in this setting that the carbonaceous sediments were laid down. As the sea continued to rise, marshes were transformed to a shallow sandy shoreline that was fed sediment by rivers that ferried the sand from the western highlands. The upper shell-rich sandstone that we now look upon was deposited in this setting. The sea, however, was not yet finished. The prolonged deepening shifted the shallow sand-dominated shoreline farther west and changed the setting here to a mud-dominated deep-sea floor. The vertical gradation from sandstone to the overlying black, organic-rich Tropic Shale confirms this event. The Tropic Shale marks the maximum sea level during the Cretaceous. Such thick accumulations of shale are not encountered elsewhere in the succession.

As we look ahead, past the outcrop of Dakota Sandstone, the valley widens considerably, a reflection of the easily eroded Tropic Shale that makes up the lower tree- and debris-covered slopes. Sandstone of the overlying Straight Cliffs Formation lines the valley for the next few miles. The Straight Cliffs is composed of four members, each representing a shift in the depositional environment and resulting lithology. In ascending order they are the Tibbet Canyon Member, the Smoky Hollow Member, the coal-rich John Henry Member, and the Drip Tank Member (fig. 12.2). We will pass through the more than 1,000-foot-thick Straight Cliffs Formation on a member by member basis.

3.8 ≈   The 60-foot cliff of yellow brown sandstone at road level is the Tibbet Canyon Member of the basal Straight Cliffs Formation. This massive marine sandstone grades up from low-energy shale of the Tropic and signals a regression of the sea so that deep-sea muds were succeeded by offshore sandbars and beach sands. Small-scale crossbedding bared in the roadcut chronicles the back and forth swash of 100-million-year-old waves as the water shallowed enough for the sea-floor sands to be molded by the currents.

4.1 ≈   The fluvial Smoky Hollow Member can be seen across the valley to the north. Its thin, ledgy cliffs and slope-forming tendencies contrast with the steep-walled Tibbet Canyon Member below. In detail, the resistant ledges are constructed of overlapping large-scale lenses of crossbedded sandstone and conglomerate representative of discrete river channels. The Smoky Hollow was deposited as the sea continued to retreat eastward. Its rivers gnawed their way through the former domain of the sea, dropping some of the sediment that was perpetually flushed from the Sevier highlands.

The first thick cliff in the middle of the slope is the coal-bearing John Henry Member. It is separated from the underlying Smoky Hollow by a subtle erosional unconformity. As the sea abandoned the area, rivers were revitalized, leading to erosion instead of deposition. Some of the upper Smoky Hollow sediments were washed out to sea at this time. During the ensuing John Henry deposition the vanquished sea apparently began a comeback, reducing the rivers' energy enough so that sedimentation again prevailed. As the sea continued to rise and step west, swamps again lined the coast, and the controversial coals of the Kaiparowits Plateau were deposited. Thinner sandstone beds associated with these coals can be seen at the top of the hill. Coal is rarely exposed in natural outcrops because it readily erodes into recesses that are then concealed beneath slope wash and debris from overlying rocks.

7.0 ≈   Cliffs that tower over the road to the south are part of the John Henry Member. A close inspection reveals a wall of stacked lenses filled with sandstone and conglomerate. Rounded quartzite and chert pebbles indicate that the Sevier highlands source was composed of older Paleozoic sedimentary rocks. Paleozoic strata were compressed into high mountains on thrust faults and large folds, submitting the formerly deeply buried rocks to the indignities of weathering and erosion and scattering their broken remnants across the vast foreland basin.

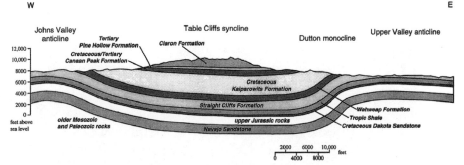

Figure 12.3. Cross section through part of the Kaiparowits Plateau just north of the route taken by State Highway 12 between the towns of Escalante and Henrieville. The high part of the Table Cliffs syncline, which exposes the Tertiary strata, is the south edge of the Table Cliff Plateau. After Lidke and Sargent 1983.

8.7 ≈   The valley once again widens as the soft Tropic Shale reappears at the surface. Strata on both sides of the valley are dipping slightly eastward; ahead, however, they tilt to the west. We are approaching the crest or axis of the Upper Valley anticline, which pushes the Tropic up from the sub-surface. This structure is one of several related folds that rippled across the Kaiparowits during the Tertiary Laramide orogeny (fig. 12.3).

     Within this part of the valley the lower sandstone cliff above the covered Tropic slopes is the shallow-marine Tibbet Canyon Member. The overlying slopes and thin cliffs are the fluvial Smoky Hollow Member. Together they record the regression of the Western Interior sea following its maximum rise, which is documented in the deep sea floor deposits of the Tropic.

10.5 ≈   The west limb of the Upper Valley anticline abruptly takes on a much steeper incline here: the strata dip as much as 25° to the west (photo 13). This pronounced asymmetry caused geologists to designate this steep limb as the Dutton monocline, named for geologist C. E. Dutton of the Powell Survey, the first geologist to study the region systematically (fig. 12.3).

13.6 ≈   The wooded hills to the south are the Wahweap and Kaiparowits Formations, the upper part of the Cretaceous succession. They are poorly exposed here as a consequence of the abundant shale and siltstone they contain. These units will be seen in much better exposures as we drive west.

     To the northwest is the first clear view of the Pink Cliffs step of the Grand Staircase, along the southern escarpment of the Table Cliffs. The southmost promontory, named Powell Point by Dutton, for Major John Wesley Powell, reaches an elevation of more than 10,000 feet. Powell

Photo 13. Tilted Cretaceous strata on the west-dipping Dutton monocline, one of several subtle Tertiary folds formed on the Kaiparowits Plateau during the Laramide orogeny.

Point was an important reference point during the initial topographical and geological surveys of the region.

The Pink Cliffs are composed of Tertiary sedimentary rocks, mostly the Claron Formation. In the trees on the lower slopes of the Table Cliffs are the earlier Tertiary rocks, the Grand Castle Formation and the overlying Pine Hollow Formation, a locally occurring unit. These are overlain by the Claron, whose pink, orange, and white strata have been carved into a crenulated rim of fins and parapets.

17.4 ≈   At this point we have crossed the drainage divide into the Paria River system and entered "the Blues," a steep badlands carved into the green-gray mudstone and sandstone of the Kaiparowits Formation. Here the formation reaches a thickness of at least 3,000 feet. This immense sequence was deposited by rivers that rolled lazily across a low, featureless floodplain to their eastern rendezvous with the sea. Contained within these low-energy deposits are the skeletons and fragments of turtles, crocodiles, dinosaurs, fish, and small mammals. Abundant petrified wood suggests that at least part of the monotony of the floodplain was broken by forests or stands of trees.

Photo 14. Concretions of more highly cemented rock weathering from sandstone beds in the Kaiparowits Formation.

What appear to be bulbous rounded boulders protruding from ledges of sandstone are not boulders, but sandstone concretions (photo 14). These elliptical or spherical concretions are simply areas that, for some reason, are better cemented than surrounding sandstone and thus more resistant to breakdown by water and wind. They are especially noticeable as we wind down the steep road that cuts the Blues.

The drab-colored Kaiparowits contrasts with the vibrant pastels of the Tertiary strata that form the backdrop for the Blues. The colorful columns of the Claron Formation that stand beneath Powell Point are the same landforms that make Bryce Canyon National Park such a unique place. Ornamentation of the Table Cliffs, however, is not yet as advanced as in the dissected wonderland of Bryce.

21.2 ≈   The cliffs that overshadow the view directly ahead, to the west, provide a spectacular mural of the Wahweap Formation, etched into sandstone and conglomerate. These high-energy deposits accumulated as rivers, choked with the debris of the disintegrating Sevier highlands, tumbled eastward.

The basal part of the Kaiparowits Formation makes up the ledgy cliffs that are stacked back from the top of the Wahweap. Black, carbon-rich

Photo 15. A narrow fin composed of the fluvial Drip Tank Member of the Cretaceous Straight Cliffs Formation. The slopes at the top are composed of the Wahweap Formation. The eastward tilt of these strata is due to their location on the east limb of the Johns Valley anticline, another product of the Tertiary Laramide orogeny.

mudstone of the Kaiparowits can be seen in the high shadowed alcoves beneath the sandstone overhangs.

23.0 ≈   The spur of tilted strata that confronts us to the south is the Drip Tank Member, the top of the Straight Cliffs Formation (photo 15). This 300-foot cliff is constructed of sandstone and conglomerate, the legacy of long-disappeared rivers that ground away at the sporadically renewed highlands to the west. The backstepping slopes above are the Wahweap Formation.

The obvious eastward tilt to these strata marks the east limb of the Johns Valley anticline, a Laramide fold noted for its strong influence on depositional patterns in overlying Tertiary strata (fig. 12.3). At the end of the Cretaceous and through much of the Tertiary this fold acted as a regional partition, disrupting the earlier single basin into several smaller ones. In addition, erosion of the anticline provided a local sediment source for the new Tertiary basins.

25.5 ≈   As the road flattens into a wide valley, we are again in the Tropic Shale. To the north the lower hills are held up by the Straight Cliffs Formation,

Photo 16. The Jurassic Entrada Sandstone unconformably overlain by the Cretaceous Dakota Sandstone near Henrieville. The lower slopes are gypsiferous shales and silt-stones capped by white eolian sandstone cliffs, both part of the Entrada. These are overlain by coarse fluvial sandstone of the Dakota, which grades upward into black, thinly bedded coals and shale. The caprock consists of fossiliferous shallow-marine sandstone at the top of the Dakota.

backed by the rest of the Cretaceous succession and, finally, the Tertiary strata of Powell Point and the Table Cliffs.

26.8 ≈   The white cliffs and pedestals of Entrada Sandstone contrast with the yellow and gray of the overlying Dakota Sandstone, which forms the base of the great Cretaceous succession (photo 16). Since our last encounter with the Dakota we have lost the Morrison Formation, which separated the Entrada and Dakota near Escalante. The Morrison is the casualty of a westward increase in pre-Dakota uplift connected to early tectonic activity in the Sevier orogenic belt. This resulted in deeper levels of erosion to the west and deposition of the Dakota on increasingly older rocks in that di-rection.

Viewed up close, pebbles and cobbles at the base of the Dakota can be seen to fill irregular hollows in the erosion surface that was scoured into the top of the Entrada. These rounded fragments are the largest yet seen in the Dakota, attesting to our increasing proximity to their ancient mountainous source. Overlying this thin basal conglomerate is a thick sequence of black and gray sandstone, shale, and coal. A closer look at these rocks reveals black mats of carbonaceous plant fragments, including recognizable stems and leaves. The top of the Dakota is defined by a continuous thin cliff of fossiliferous marine sandstone.

28.2 ≈   To the south white and red siltstone in the middle part of the Entrada is exposed along the lower slopes.

29.0 ≈   We enter the Henrieville town limits (sign). This small ranching community was first settled along the banks of Henrieville Creek in 1878 by a group of Mormons from nearby Cannonville. The creek and the quiet little town were named for James Henrie, one of the early settlers.

29.8 ≈   On the west edge of Henrieville the view opens in all directions to reveal a wide encirclement of cliffs and benches. To the north the low white and red-brown slopes are interbedded siltstones and sandstones of the middle Entrada. The fluted white columns above are eolian sandstone of the upper Entrada. The highest parts of the sequence are the various facies of the Cretaceous Dakota Sandstone.

To the south deeper levels of erosion have exhumed the red slickrock benches of the lower Entrada. This slickrock is composed of red, large-scale crossbedded sandstone of obvious eolian origin. Thin interbeds of purple and white siltstone probably mark small incursions of the shallow Sundance sea that made episodic forays south into the area.

30.9 ≈   Directly ahead, on the skyline, lies the forested southeast edge of the Paunsaugunt Plateau, a part of the High Plateaus section of the Colorado Plateau that separates it from the Basin and Range Province to the west. Exposed along the plateau margin are the colorful Tertiary rocks of the Claron Formation. We are looking at Bryce Canyon National Park. Concealed between us and the plateau is the north-trending Paunsaugunt fault, one of several major north-trending faults with a down-to-the-west sense of movement that mark a broad transition zone between the Colorado Plateau and the Basin and Range. This fault will be treated in more detail as we approach it on our way to Bryce.

31.8 ≈   The steep roadcut in this "pass" provides a better glimpse of the silty part of the Entrada. These crumbling cliffs are lined with interbedded siltstone and sandstone, with the more resistant sandstone standing out in relief relative to the siltstone recesses. Sandstone units show horizontal and ripple laminations that depict the coming and going of tidal currents in the shallow, ephemeral sea that lay here 165 million years ago.

32.5 ≈   We enter the town limits of Cannonville (sign), one of several closely spaced settlements in the upper Paria River valley. Like most of the small towns in southern Utah, it was settled by Mormon pioneers; the first families arrived in 1854. Originally the town was called Clifton for the picturesque cliffs of Jurassic strata that practically surround it. The name was later changed to Cannonville, for a prominent church official. It is assumed that the settlers had weightier matters on their minds than coming up with fanciful names for their villages.

32.6 ≈ The road crosses the Paria River. Most of the time it is a stretch of the imagination to call it a river: for three-quarters of a typical year it scarcely holds a trickle. The Paria winds south through a rugged, deeply incised terrain before joining the Colorado River at Lee's Ferry, where the Grand Canyon begins. Headwaters of the Paria are less than 10 miles north in what is known as the Paria amphitheater. This great bowl is encompassed by small streams that emanate from the surrounding escarpments of the Table Cliffs and Paunsaugunt Plateau to feed the Paria. It is only in the spring, however, that these tumbling creeks possess any vigor.

32.8 ≈ Highway 12 curves north to Tropic and Bryce Canyon; the road that turns south goes to Kodachrome Basin State Park, where the rich-hued Carmel Formation and Entrada Sandstone have been sculpted into a dazzling assortment of cliffs, fins, and pillars. This graded dirt road continues south along the Cockscomb, where it winds through fantastic fins of upturned sandstone. The Cockscomb, which is the northern continuation of the lengthy Kaibab monocline, defines the western edge of the Kaiparowits Plateau. After about 50 miles the road deviates from the trace of the monocline to emerge into a vast gray badland of Tropic Shale. About 10 miles farther south the dirt road reaches U.S. Highway 89, which can be taken east to Glen Canyon dam or west to Kanab. This remote road traverses a beautiful and seldom-visited part of Grand Staircase–Escalante National Monument. However, you should be prepared by having plenty of gas and water. If there has been any rain or snow this road should not be used, because the Tropic Shale turns into bottomless gumbo when wet. You have been warned!

34.0 ≈ The towers that crowd the road on the west are composed of the middle siltstone and upper sandstone of the Entrada, capped by several feet of the basal Dakota conglomerate. The upper eolian Entrada sandstone is much thinner here than just a few miles east, apparently the victim of pre-Dakota beveling. To the east lies the Paria River floodplain, currently utilized for grazing and hayfields. The active part of the river channel is well below the floodplain level, incised into the alluvium that fills the valley.

34.5 ≈ We are now in the Dakota Sandstone. The basal fluvial part is up to 30 feet thick here, a striking change from about 5 feet near Henrieville and its absence near Escalante. These variations reflect localized differences in the depth of pre-Dakota incision. Dakota rivers did not flow across a planar surface and, in fact, were initially confined to channels cut into underlying rocks, in a manner reminiscent of modern streams on the Colorado Plateau. For example, if today's canyons and valleys were to slowly fill with sand and gravel prior to inundation by a rising sea, dramatic variations in thickness would be preserved. Pre-Dakota topography, however, was not nearly as pronounced as that of the modern Colorado Plateau.

35.7 ≈   The roadcut on the west side is in the upper marine part of the Dakota. These thin, horizontal-bedded sandstone units alternate with black shale, suggesting a deep-sea environment. Horizontal bedding indicates that sediment settled from suspension in water too deep to be affected by normal wave or storm currents.

37.2 ≈   As we enter the town of Tropic, the pink cliffs of Bryce can be seen up the draw to the west. The town is surrounded, appropriately, by gray slopes of Tropic Shale. This is its type locality, the place for which the formation was formally defined and named. The first Mormon settlers to enter the valley from Panguitch, seeking a milder climate, chose the name, inspired by the lower elevation and warmer climate of the upper Paria River valley.

38.0 ≈   Blocky sandstone beds above the Tropic are the lower part of the Straight Cliffs Formation, the equivalent of the Tibbet Canyon or Smoky Hollow Member to the east.

40.4 ≈   As we climb through Tropic Canyon, this deep notch in the flank of the Paunsaugunt Plateau, the trace of the Paunsaugunt fault can be seen in a gully to the north-northwest. The gray Cretaceous rocks on the east side are juxtaposed with the orange and pink Claron Formation to the west. Offset on this part of the fault is estimated to be 1,200 feet, but over its total length of at least 120 miles the amount of displacement varies. Movement is thought to have occurred about 8 to 5 Ma. We have been traveling parallel to and east of the fault since Cannonville, but its path in the hillsides is obscured by alluvium and trees. This north-trending fault bounds the east edge of its namesake plateau, yet movement was very different than the present landscape would suggest. The west (plateau) side of the fault has moved *down* relative to the east (Paria River valley) side (fig. 12.4). If this is so, why does the present topography suggest just the opposite? The answer lies in different rates of erosion for the two sides of the fault, controlled by such factors as relative elevations and the rock types involved.

Prior to faulting, the Claron blanketed the region. As the extensional stress that had been cutting the Basin and Range Province with faults encroached on the west margin of the stable Colorado Plateau, several large-scale normal faults, including the Paunsaugunt fault, developed. When the west side of the Paunsaugunt fault dropped, erosion attacked the relatively

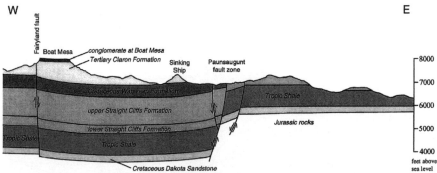

Figure 12.4. Cross section across the Paunsaugunt fault in Bryce Canyon National Park. The section is located approximately south of State Highway 12. Note that the Paunsaugunt fault is actually a fault zone here, in contrast to the single discrete fault where Highway 12 crosses it. After Bowers 1990.

high east side (fig. 12.5). Eventually the Claron on this side was stripped away, revealing the soft Cretaceous rocks beneath. At this time creeks and streams continued to cut into the easily eroded Cretaceous strata, leaving the Claron on the west side unscathed. As the Cretaceous rocks were ground away at an accelerated pace, the east block was transformed to a deep valley, while the resistant west side became a high plateau.

Even as this drainage pattern was being established, headward erosion bit into the flanks of the plateau, dissecting the Claron into a myriad of shapes. Erosion of the Claron, however, was not a straightforward affair. Stresses along the fault during movement were transmitted into the adjacent Claron, lacing the brittle limestone with fractures and joints long before it was unearthed by erosion. It is these fracture patterns that control the orientation and geometry of the fins and pinnacles that define the landscape of Bryce Canyon. The bizarre quality of the landforms is intensified by their vertical undulations, caused by the different weathering rates in the alternating limestone and siltstone beds.

The Paunsaugunt fault is one of three regional, north-trending normal faults that characterize the High Plateaus section of the Colorado Plateau. All the faults accommodate large amounts of down-to-the-west movement and define the boundaries of the individual plateaus. The Paunsaugunt fault, as we have seen, marks the eastern limit of these faults and separates the Aquarius Plateau/Table Cliffs region from the Paunsaugunt Plateau (fig. 12.6). As we move west, the Sevier fault outlines the western edge of the Paunsaugunt Plateau by forming the Sevier River valley, which separates the Paunsaugunt from the Markagunt Plateau to the west. The west margin of the Markagunt is bounded by the escarpment of the Hurricane fault, which is the major geologic boundary between the Colorado Plateau

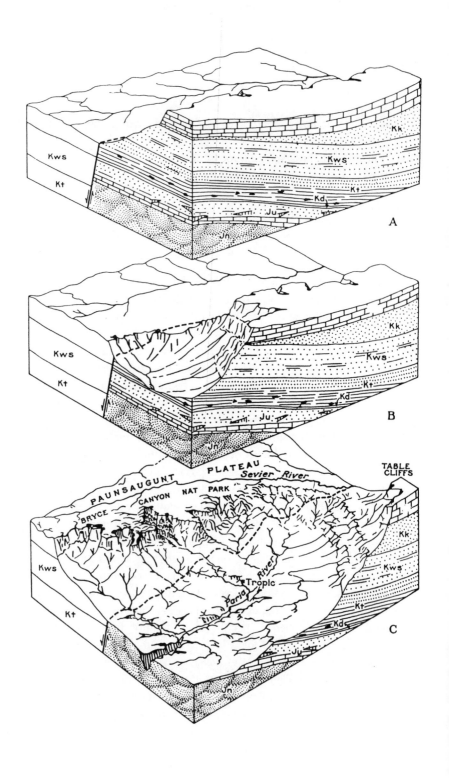

A

B

PLATEAU
PAUNSAUGUNT   *Sevier River*   TABLE
CANYON NAT PARK   CLIFFS
BRYCE

Kk

Kws

Tropic
River

Kt

Paria

Kd

Ju

Jn

C

Kk

Kws

Kt

Kd

Ju

Jn

Kws

Kt

and the Basin and Range. While each fault steps downward to the west to create a stair-step transition zone between the disparate crustal blocks, it is the Hurricane fault that truly isolates the flat-lying sedimentary strata of the stable Colorado Plateau from the tilted, faulted, and folded igneous and sedimentary rocks of the Basin and Range.

41.2 ≈   The first stop after crossing the fault and entering the brilliant world of the Claron Formation is the trailhead to Mossy Cave on the south. Here the orange, white, and pink rocks that make up Bryce can be examined up close.

The canyon that the trail climbs through is unusual. Abundant water cascades through it year round, yet most of the drainages off this east flank of the plateau are dry. The flow of this creek has been enhanced by the hard labor and sweat of the Mormon pioneers. In the early 1800s, after realizing that the Paria River was an unreliable water supply, the settlers of the upper Paria River valley began to construct a ditch across the Paunsaugunt Plateau to divert water from the East Fork of the Sevier River, which flows north to bisect the vast tableland. After several years of toil the ditch was completed, and water was redirected into the head of the natural drainage before us, which was called Water Canyon. Now it tumbles down this steep canyon, but before meeting the gravel floor of Tropic Canyon it is once again restrained by a ditch, where it is forced into a course mandated by pioneers to feed their parched fields. This is one of the few instances where water destined for the oblivion of the Great Basin has been rerouted onto the Colorado Plateau.

The Claron Formation, which ranges from Late Paleocene to Eocene

Figure 12.5. (*opposite*) A series of diagrams showing the evolution of the upper Paria valley and the west margin of the Paunsaugunt Plateau with respect to movement on the Paunsaugunt normal fault. *A* shows the relative uplift of the east side of the fault, raising the Tertiary Claron Formation on that side well above the modern Paunsaugunt Plateau surface. *B* shows the preferential erosion of the high east side of the fault, removing the hard caprock of the Claron to expose the soft underlying Cretaceous rocks. At this time the exposure of less-resistant Cretaceous rocks on the east side caused it to be eroded lower than the downthrown west side, which retained its resistant caprock of the Claron Formation. *C* shows the modern topography and the trace of the fault. The landscape of Bryce Canyon National Park formed as the side canyons cut headward into the east flank of the Paunsaugunt Plateau. Rock units: Jn, Jurassic Navajo Sandstone; Ju, Upper Jurassic Entrada Sandstone; Kd, Cretaceous Dakota Sandstone; Kt, Tropic Shale; Kws, Wahweap and Straight Cliffs Sandstone; Kk, Kaiparowits Formation; the Tertiary Claron Formation is shown in a brick pattern on the east side of the fault and is the unlabeled unit on top of Kws on the west side of the fault. After Gregory and Moore 1931.

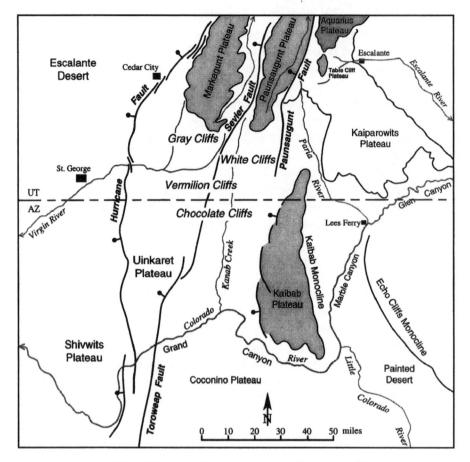

Figure 12.6. Map of the High Plateaus and the southwest part of the Colorado Plateau showing the major normal faults formed during Basin and Range extension and some of the monoclines that formed earlier during the Laramide orogeny. The Colorado Plateau is defined geologically as being east of the Hurricane fault, while the Basin and Range Province is west of the fault.

age, consists dominantly of limestone with minor amounts of silt and clay and rare sandstone and conglomerate. These rocks were deposited in a large closed basin that probably encompassed most of southwest Utah. The basin was bordered to the east by the Circle Cliffs uplift, with the west margin defined by the highlands of the Sevier orogenic belt, which had been inactive for at least 5 million years. Over most of its history the basin was filled by a vast shallow lake in which calcium carbonate accumulated from both chemical and biogenic processes. Sparse clastic sediment was supplied by rivers that originated in the highlands to the west. On the west margin of the Paunsaugunt Plateau, in the Red Canyon area, conglomerate lenses become increasingly conspicuous and probably represent the preserved mouths or channels of these rivers.

The most striking feature of the Claron is its luminescent color, the product of varying trace amounts of iron oxide in the limestone. The only consistent sedimentary structures in the formation are crude horizontal lines that separate different beds. Finer features such as crossbedding or laminations are indistinct. Fossils of freshwater gastropods, however, turn up occasionally.

The obscuring of structures that at one time most certainly existed is attributed to soil-forming processes. As the water level fluctuated, lake-floor sediments became exposed to weathering and chemical alteration. Intense soil development in preexisting sediments typically diminishes all but the largest structures, in this case leaving only the bedding planes that define changes in rock type.

42.6 ≈   In the middle of the orange and white cliffs that watch over the road to the northwest is one of the most enigmatic structures encountered so far, the Rubys Inn thrust fault. The fault trace is marked by the recessed area that undulates across the middle of the cliff face. The black-streaked overhang seen from this view is the grooved fault surface. The rocks involved in this segment of the thrust are the lower Claron thrusted over the lower Claron, repeating the formation. This relationship makes the fault difficult to recognize.

The Rubys Inn thrust is puzzling for two reasons. First, the movement was southeast-directed, indicating southeast-northwest–oriented compressional stress. This orientation fits none of the well-documented orogenies that have affected the region. Second, and most peculiar, is the timing of the compressional event. The last documented event was the Laramide orogeny, which was over by about 50 Ma, while Claron sediments were still accumulating. Because the Claron is involved in the deformation, faulting could not have occurred any earlier than the Late Eocene, at about 40 Ma. It is possible that some small-scale post-Laramide pulse of compression rippled through the area, but no other faults or folds of this orientation or age have been recognized elsewhere in the region. There is no simple "textbook" explanation for the Rubys Inn thrust. We know the geometry and have some constraints on its age (e.g., < 40 Ma), but the cause remains a mystery. It is somehow reassuring to know that such features remain in this great geological laboratory of the Colorado Plateau.

43.9 ≈   From the top of the Paunsaugunt Plateau, a broad panorama unfolds. The plateau summit is covered by open grassy meadows fringed with stands of aspen and towering ponderosa pine, which in the midsummer offer a soothing respite from the blazing heat of the lower valleys. The name "Paunsaugunt" comes from the Paiute Indians who used to roam this land and means "home of the beaver." It would probably be easier to see why if we could go back 150 years, before the water diversions, grazing, and constant traffic of Bryce Canyon that now disrupt the natural setting.

Over most of its extent the Paunsaugunt is capped with the Claron Formation. Forested hills to the west and southwest are composed of the upper Claron, which in the rim area has been stripped back. To the north the landscape takes a 3,000-foot step up into the volcanic rocks of the Sevier Plateau. In reality the Sevier is simply a higher, northern part of the Paunsaugunt Plateau, but the abrupt rise in elevation and profoundly different rock type legitimize its differentiation, at least in the eyes of geologists. Its craggy cliffs and rugged mountains are composed of the Needles Range and the Mount Dutton Formations, both the product of large-scale volcanism.

Rocks of the Needles Range Formation rest on the Claron and consist of welded tuff that is spread over a large part of southwest Utah. Welded tuff forms during explosive volcanic eruptions when hot volcanic ash and debris is violently blown from the central vent. After falling from the air or flowing laterally as an incandescent cloud of ash and boiling gas, the hot particles settle and "weld" together to form a dense volcanic rock. The age of the Needles Range Formation spans about 31 to 26 Ma. Although it covers an immense area, the location of the eruptive center remains unknown. It probably is hidden beneath a blanket of younger volcanic rocks or has been down-faulted in the Basin and Range to the west.

The voluminous Mount Dutton Formation is the product of the Marysvale volcanic field and comprises most of the Sevier Plateau rocks. The formation was defined by J. J. Anderson, P. D. Rowley, and others in 1975 and was named for Mt. Dutton, which at 11,040 feet is the highest point on the Sevier Plateau. For the most part the Marysvale volcanic field was a single stratovolcano with a diameter of more than 30 miles, closely resembling the modern volcanoes of the Cascade Mountains in the Pacific Northwest. The formation consists of a central eruptive part, mostly andesite and basalt lava flows, and an outlying area dominated by debris flow deposits that were jettisoned from the flanks of the volcano during frequent periods of instability. The age of the Mount Dutton rocks ranges from about 26 to 21 Ma.

The 30 to 20 Ma time span for the eruption of the Sevier Plateau volcanic succession puts it into the time frame of the laccoliths of the Henry Mountains and the Abajo and La Sal Mountains, as well as the gigantic

San Juan volcanic field in southwest Colorado. This regional igneous flare-up is part of the post-Laramide magmatism that swept across the western United States.

44.8 ≈   In the high, clear air of the plateau we reach a crossroads. To continue west on State Highway 12 would soon take us down off the west edge of the Paunsaugunt into the Sevier River valley, where this road ends at U.S. Highway 89. If 89 is taken south, it eventually leads to Zion National Park and/or the town of Kanab. If we turn north at this intersection, State Highway 22 rolls along the East Fork of the Sevier River, through Johns Valley and the settlements of Antimony and Koosharem. This route is rimmed on the west by the volcanic cliffs of the Sevier Plateau and on the east by the less dramatic slopes of the Awapa Plateau. The road to the south takes us past Rubys Inn and into Bryce Canyon National Park. This road, which will constitute the next segment of the road log, dead ends at Rainbow Point in 20 miles.

13

Bryce Canyon National Park

mile
0.0 ≈   At the crossroads we turn south onto State Highway 63, which takes us to Rubys Inn and Bryce Canyon National Park. For this segment of the road log only a few of the dozen overlooks in Bryce will be treated in detail. A beautiful picture book that concentrates exclusively on the geology of Bryce called *Shadow of Time* by Frank DeCourten is available at the park visitor center.

1.3 ≈   On our way to the park entrance we must first negotiate the congestion of Rubys Inn, part of that great cluster and clutter of trinket stands, hotels, restaurants, and throngs of smoky buses that plague National Park boundaries throughout the country.

After we pass through the park entrance, the visitor center, with trail maps and books, is to the west. It is a good place to get properly oriented. While most of the overlooks are crowded with tourists, many of the trails that wind down through the colorful pinnacles and arches are relatively quiet. Some trails are more crowded than others, and the visitor center is a good place to inquire about such things.

Sunset Point ≈

4.8 ≈   We turn east to reach Sunset Point. The view from here is roughly similar to that from Sunrise, Inspiration, and Bryce Points. Sunset Point stands at the head of the fantastically crenulated amphitheater of Bryce Canyon. The ephemeral waters of Bryce Creek spill from the rim that encompasses this rugged bowl, down the gullied maze and eastward off the flank of the plateau. This seasonal cycle of flowing water works today as it has for millions of years.

The sculpting responsible for this army of glowing hoodoos and serrated fins is never-ending. It began with fractures that permeated the brittle Claron rocks to form a crisscrossing network. According to geologist

William Bowers (1990) of the U.S. Geological Survey, major joints in Bryce form two distinctly oriented sets. In map view these main joints trend N 33° W (33° west of due north) and N 21° E (21° east of due north). These fractures probably were generated by one or both of the recognized tectonic events young enough to have affected the Claron, extensional stress associated with the Paunsaugunt fault and compression during thrusting along the Rubys Inn fault.

As water trickled into fractures at the surface, freezing temperatures periodically turned it to ice, expanding and forcing the fractures wider. Even today the climatic setting of the park makes this freeze-thaw action an important agent in the breakdown of the rocks. Eventually the cracks grew into passages through which flowing water was focused; particle by particle the sediment was dislodged and removed.

As sediment was removed from the lofty plateau and transported to some lower destination, the journey was a deliberate one. Upon reaching the Paria River in the valley below, the sand grains began a methodical descent to the Colorado River, eventually to be washed to the Colorado River delta in the Sea of Cortez, far to the south. Anywhere along this path the grains may have paused for a geologic moment, sheltered from the currents between cobbles on the river floor or stored high on a floodplain or sandbar. But this was always temporary. Even today huge volumes of sediment lie waiting in Lake Mead to resume their seaward journey. Now the stagnant delta receives no sediment and very little water. Instead it waits patiently for the moment when the concrete plugs are breached— an imminent event, geologically speaking.

The vertical convolutions of the spires and fins reflect variations in the rock types that were laid down in the lake deposits of the Claron. The wider, bulging layers are composed of the resistant carbonate rocks limestone and dolomite. In wetter regions such as the eastern United States carbonate rocks are readily dissolved to form valleys and low-lying areas. In the arid West, however, they are the most resistant of the sedimentary rocks and often form a wide-brimmed caprock that shelters weaker underlying rocks from the elements. In contrast, siltier, clay-rich rocks are easily eroded and, in Bryce, tend to form recesses, alcoves, and narrow, fragile-looking horizons. Many of the windows and arches that pock the narrow fins formed in the clay-rich layers. Sandstone, which is present sporadically in the Claron at Bryce, lies somewhere between carbonate rock and shale in its resistance to erosion. All these rock types contribute to the unique landscape that now confronts us.

The electrifying colors of the Claron Formation are its most eye-catching feature. The glowing tints of red, orange, and pink are caused by trace amounts of iron oxide or hematite, as it is known to geologists. Any time that iron is exposed to oxygen, through the atmosphere or water, the two common elements readily combine into iron oxide, which is rust. The phosphorescent white layers in the Claron are simply devoid of hematite.

Looking eastward to the town of Tropic, our eyes cross the Paunsaugunt fault. It is not recognizable, however, for here it places Cretaceous rocks against Cretaceous rocks so that no revealing color change marks its trace. Perhaps better evidence lies across the Paria River valley, along the buttress of Table Cliff Plateau. If we use the informal pink limestone member of the Claron Formation as a reference point, changes in elevation across the valley have some important implications. Just below the promontory of Powell Point on the eastern skyline, the top of the pink limestone sits at an elevation of 9,840 feet. In contrast, Sunset Point stands at the top of the pink limestone, but at an elevation of 8,000 feet. Thus we are provided with an approximation of the amount of movement across this fault, with our (west) side dropping 1,840 feet!

Paria View ≈

5.2 ≈   A short distance south on the main road another turnoff leads to three more overlooks. We turn here and bear right to Paria View. The other two, Inspiration Point and Bryce Point, look northward into the Bryce amphitheater, while Paria View provides an unimpeded view to the southeast. After a short walk to the overlook we are rewarded with a vista that includes three of the five steps of the Grand Staircase, a phrase coined by Major John Wesley Powell upon viewing this area from the south. We are perched on the top step, the Pink Cliffs, composed of the colorful Claron Formation and carpeted with ponderosa pine and stands of delicate aspen. One step down are the less distinct Gray Cliffs, composed of gray and brown Cretaceous strata. In reality, the broken-up Gray Cliffs resemble several small steps: they are divided into a series of brown sandstone cliffs and gray piñon-covered shale slopes. Still farther south and another step down are the White Cliffs, defined by the impassable Navajo Sandstone, but capped by the Carmel Formation.

Immediately below us the open grassy meadow next to the bed of Yellow Creek is maintained by a spring that emits from the Peekaboo fault, a north-trending normal fault. This fault lies less than 2 miles west of the similarly oriented Paunsaugunt fault and probably is related to it. Like the Paunsaugunt, the Peekaboo shows down-to-the-west movement, but the offset is small. Springs commonly issue from faults. When fault movement takes place, rock in the fault zone may be pulverized, creating a vertical conduit of highly porous rock. This may then become a passageway to the surface for groundwater that otherwise is confined to porous rock layers deep beneath the surface.

We now proceed to the south end of the park. Along the way we pass numerous overlooks and trailheads, all with unique and multicolored views. Below us, amid the kaleidoscopic hoodoos, winds the Under-the-Rim trail, which extends from Bryce Point southward to Yovimpa Point, a distance of 10 miles as the crow flies, but easily twice that as the backpacker walks. The trail can be conveniently accessed from most of the trailheads and overlooks along the road, however, so that shorter segments can be explored. I have rarely seen people on this trail, even after descending from crowded, noisy overlooks.

Rainbow and Yovimpa Points ≈

20.4 ≈   Approaching Rainbow Point, we have reached the end of the road and the southernmost vistas in the park. At 9,115 feet we stand at one of the higher parts of the park on an island composed of the upper white member of the Claron, a unit that has been absent or concealed by trees until now. Beneath this isolated white mesa stand the ragged orange turrets and shadowed alcoves of the dissected pink member. The brightly colored slopes and soils at their feet offer a glimpse into the future of these proud standing rocks, for that ultimately is their fate. Looking northeast into the green blanket of the trees, we see tan sandstone buttresses of the Wahweap and Straight Cliffs Formations, seemingly holding the low gray ridges aloft. An interesting aspect of the Cretaceous succession from here is the absence of the Kaiparowits Formation, which normally is present between the Wahweap and the overlying Claron Formation. We are looking up the north-trending axis of the Bryce Canyon anticline, a subtle product of the Laramide orogeny. This gentle uparching put the Kaiparowits Formation in an exposed position at the crest of the fold as it was beveled prior to Claron deposition. The gentle east and west dip of the Cretaceous sandstones is barely perceptible from this vantage point. As our eyes drift farther, toward the Paria valley, the bright red Upper Jurassic rocks that surround the Henrieville area can be seen as they again give way to the Cretaceous and Tertiary rocks that lead upward to the Kaiparowits and Table Cliff plateaus.

A short hike to Yovimpa Point provides unparalleled views off the southern escarpment of the Paunsaugunt Plateau. While the Claron has a rather subdued expression, it is the more distant horizons that make this overlook exceptional. The tree-covered slopes and low mounds below are the Gray Cliffs of Cretaceous strata. Here the Gray Cliffs step is an unimpressive blanket of vegetation, although vegetation in this arid land is in itself a wonder. Beyond that loom the great White Cliffs of Navajo Sandstone. The white, north-facing escarpment is part of No Mans Mesa, a remote island of Navajo Sandstone completely encircled by deep canyons. The pointed summit south of the mesa is Mollies Nipple, another erosional remnant of the Navajo isolated by deeply incised drainages. On the distant southeast horizon the broad hump of the Kaibab Plateau can be seen. This high plateau is pushed up on its east flank by the Kaibab monocline, one of the more extensive monoclinal flexures on the Colorado Plateau. The Colorado River slices this massive uplift to create the gorge of the Grand Canyon. Farther east, if the haze permits, we can see the dome of Navajo Mountain, an isolated igneous intrusion mantled by sandstone. The dome is believed to be the same age as its igneous neighbors, the Henry and Abajo Mountains to the north. The great hulking mass is sacred to the people of the Navajo tribe, and its flanks host some of the most beautiful and mysterious canyons on the Colorado Plateau.

We are at the south end of the Paunsaugunt Plateau, at the headwaters of the East Fork of the Sevier River, the main drainage system for this flat-topped highland. The East Fork flows north up the middle of the plateau from this slightly elevated south rim. Along the way small tributaries feed it from the east and west. The river continues north along the east edge of the Sevier Plateau until cutting west to join the main stem of the Sevier River.

14

"Bryce Junction" to Mt. Carmel Junction
via State Highway 12 and U.S. Highway 89

mile

0.0 ≈ As we travel west on Highway 12, the road remains in the Claron Formation. No other rock units will be encountered until we drop off the west edge of the Paunsaugunt Plateau.

2.0 ≈ We cross the East Fork of the Sevier River. For most of the year it is only a trickle, probably owing to its diversion upstream into the Tropic ditch, where it is funneled down the east escarpment of the plateau through Water Canyon then back into a ditch to be distributed to the inhabitants of the Paria valley. What is left of the East Fork merges with the Sevier River, appropriately, near the town of Junction. From there the Sevier River flows into the Sevier Desert to the west, where it evaporates or sinks into the subsurface.

6.7 ≈ As we approach the plateau's edge, we drop into Red Canyon; the landscape, in dazzling shades of red, becomes distinctly Bryce-like. The hoodoos and fluted canyon walls enable us to enjoy these canyons without the constraints imposed by the National Park system. This area is laced with hiking and biking trails and dirt roads on which to view the grotesque forms more closely.

8.8 ≈ The road passes through two short tunnels blasted through red fins of Claron limestone.

10.3 ≈ While the fins and minarets of the Claron dominate the roadside, several geologically significant features appear high on the canyon walls. Looking west from this large pullout, we see black basalt atop the red and white hill of Claron. This basalt flow is part of the suite of "young" basaltic eruptions in southern Utah that are less than 10 million years old. Lava that fed this flow may have escaped the depths of the earth through the Sevier fault, which lies immediately west of the hill.

Lower on these cliffs several gray blocky cliff bands stand out among the glowing red and white walls. These gray ribbons are constructed of coarse conglomerate; its rounded, tightly packed cobbles are testimony to the passing of tumultuous rivers on their journey to the late great Tertiary lake. Coarse clastic sediment was absent from the lake-center deposits of the Claron in Bryce Canyon. Here at the west edge of the Paunsaugunt Plateau, however, we near the northwest margin of the ancient lake. Conglomerate ribbons mark the place where the southeast-flowing rivers laid their coarse sediment to rest, as water and finer clays continued to feed into the lake.

10.8 ≈   As the canyon opens onto the Sevier River valley, the red rock of the Claron abruptly disappears. We have reached the Sevier fault, a regional-scale, north-trending normal fault that defines the west margin of the Paunsaugunt Plateau. The down-to-the-west movement on the fault has dropped the Claron below the surface of the valley that spreads before us. As we look north from the rest area, the fault escarpment is easily identified by black basalt dropped on the west side against the bright red Claron to the east.

The Sevier fault is the middle of three major north-trending normal faults that accommodate a westward down-stepping off the Colorado Plateau and into the Basin and Range tectonic province (fig. 12.6). Here at the mouth of Red Canyon the west side has been dropped 900 feet, although along the 200-mile stretch of the fault the amount of movement varies from as little as 100 feet to as much as 2,000 feet.

13.3 ≈   We cross the north-flowing Sevier River. Its headwaters lie to the south along the east flank of the Paunsaugunt Plateau and the Markagunt Plateau, which lies to the west.

13.4 ≈   At the intersection, State Highway 12 ends. We will take U.S. Highway 89 south toward the town of Kanab along the Utah-Arizona border and Zion National Park. The road follows the Sevier River upstream, almost to its headwaters, and crosses a divide into the Virgin River watershed (fig. 12.1). For the duration of our drive south on 89 the Sevier fault will remain immediately to the east. Over most of the route the fault is evident, although occasionally it is obscured or the road deviates from its trace. The following road log will track the fault until Mt. Carmel Junction, where we diverge from it to travel west to Zion.

16.0 ≈   Poorly cemented conglomerate and sandstone in the roadcut are part of the Sevier River Formation, named for the upper Sevier River valley through which we now pass. Although these sediments contain no age-diagnostic material, related lake deposits elsewhere contain microfossils and freshwater gastropods that suggest a Late Pliocene to Early Pleistocene age, in real numbers ranging from about 3.5 to 1.0 million years.

Little is known about the formation except that similar deposits apparently were widely distributed in southwest Utah and adjacent Nevada.

They now occur in isolated patches, making correlation and reconstruction of the paleogeography a difficult task. In the Sevier River valley the volcanic composition and angularity of the fragments suggest local derivation, probably from the high volcanic plateaus that encompass the region. While the exact setting remains unknown, it has been speculated that the sediments accumulated in low-lying areas that developed as extensional faulting related to the Basin and Range progressed. Their spotty distribution today is due to subsequent erosion.

18.3 ≈   Black basalt lines the road. This lava flow originates on the Markagunt Plateau, which bounds the Sevier valley on the west and is part of several very young flows in the area that may be less than 1 million years old. It has been suggested by some geologists that similar young basalts on the west side of the Markagunt escaped from vents along the Hurricane fault and that their eruption was possibly induced by fault movement. The Hurricane fault escarpment forms the west margin of the Markagunt Plateau.

21.0 ≈   We enter the town of Hatch, which was relocated to this site around 1900. Its earlier incarnation, Hatchtown, was forced from its previous location 2 miles south, where Mammoth Creek comes off the Markagunt to join the Sevier River. Flooding problems forced the inhabitants of Hatchtown and surrounding areas to higher ground. This slightly elevated terrace was chosen to make their new start.

As we continue south out of Hatch, the west scarp of the Paunsaugunt Plateau maintains its presence along the east side of the valley. Looking southeast across the pastures and hayfields, we see the Claron on the floor of the valley. The abrupt rise of steep, tree-covered slopes marks the Sevier fault, as the upper part of the Cretaceous succession is exhumed on the upthrown east side. The red Claron cliffs crown the sequence and inspire the name given to the escarpment, Sunset Cliffs. This view also emphasizes the magnitude of movement along the fault: the rocks that make up the valley floor are the same as those that cap the plateau 1,500 feet above.

29.2 ≈   The low hills along the road are in the Claron Formation, where the headwaters of the Sevier River have incised this downdropped block.

33.3 ≈   At Long Valley Junction, State Highway 14 heads west to Cedar Breaks, a less-developed version of Bryce Canyon on the neighboring Markagunt Plateau. This road continues west, to drop off the precipitous Hurricane fault scarp, and eventually to Cedar City.

The pass we are crossing acts as a high-altitude bridge that links the formidable southern ends of the Paunsaugunt and Markagunt Plateaus. It is also a significant drainage divide. As we continue south over the crest, the Sevier River drainage is left behind, and we enter the headwaters of the Virgin River, which eventually ends up in the impoundment of Lake Mead.

40.8 ≈   As we drop down the hill, basalt lines the roadcut to the east and caps the hill directly ahead. The fissure through which this particular lava flow erupted appears to be 2 miles to the east, along the Sevier fault. Continuing east down the East Fork of the Virgin River, we have cut through the Claron to enter the gray world of the Cretaceous Period. Crossbedded sandstone beds are part of the Kaiparowits and Wahweap Formations. The red and orange crenulations of the Tertiary sediments can still be seen high to the east and west, but as we continue our descent they will be left behind. Although it is concealed behind the hills, the single strand of the Sevier fault that we have been following bifurcates into a veinlike network of several smaller faults to create a diffuse *fault zone*.

42.9 ≈   Crossbedded sandstone and conglomerate on the west side of the road are fluvial deposits of the Wahweap Formation. These large Cretaceous rivers flowed east out of the highlands of the Sevier orogenic belt that lay along the present-day Utah-Nevada border.

46.9 ≈   As we approach the town of Glendale, the canyon widens to host a patchwork of irrigated hayfields and orchards. This widening canyon, like so many others we have passed through so far, marks another rendezvous with the Cretaceous Tropic Shale; its easily eroded nature again provides an ideal flat-bottomed canyon for settlement. The steep bands of thick sandstone on the western rim of the canyon are the Straight Cliffs Formation. It is thinner here than on the Kaiparowits Plateau and not as easily divided into its various members.

Part of the Sevier fault zone forms the east canyon rim. Here it branches into at least six parallel faults that sinuously cut through the area. Individually each fault accommodates only minor down-to-the-west movement, but collectively the zone accounts for about 1,800 feet of offset, juxtaposing the Jurassic Carmel Formation on the east side with the Cretaceous Wahweap Formation across the zone, approximately 1.5 miles to the west.

48.4 ≈   The white crossbedded Navajo Sandstone can be spied through the trees low on the east side of the canyon. The Navajo is on one of the small upthrown blocks caught in the fault zone. The canyon between Glendale and Orderville is cut by a complex web of faults that essentially place various Cretaceous units on the west side against the Jurassic rocks of the Navajo and Carmel Formation on the east side.

49.7 ≈   We enter the town of Orderville. This quiet little town has an interesting history. Its name derives from the United Order, a group of Mormons dedicated to the equal distribution of work and wealth through communal living. Orderville was established with just those intentions. Although attempted in other communities without much success, this exercise in self-sufficiency was surprisingly long-lived for the dedicated people of Orderville, lasting from 1875 to the late 1880s before failing for a variety of reasons. Today this neat and well-maintained town is a legacy of this bold experiment.

51.3 ≈   We cross the East Fork Virgin River in downtown Orderville. If not for this river and its constant flow, none of the towns in this valley would be viable. The East Fork supplies the inhabitants with reliable irrigation water and municipal water supplies. Even groundwater pumped from wells owes its existence to recharge by the downward seepage from the riverbed and irrigation ditches.

51.9 ≈   A spectacular show of the White Cliffs opens to the southeast. The grandiose white temples of Navajo Sandstone are overlain by the receding ledges of the Carmel Formation. We travel through the Cretaceous Dakota Sandstone and Tropic Shale, yet the older Navajo and Carmel rocks loom far above us to the east. It is the Sevier fault, obviously, that slashes between the road and these jointed cliffs, dropping the Cretaceous rocks far below their rightful stratigraphic position on top of the Carmel. Here the fault zone that we have been crisscrossing unites into a single large fault.

53.7 ≈   We enter the town of Mt. Carmel, originally called Winsor. The town was founded when settlers from surrounding areas moved into the valley with livestock and nursery stock to establish orchards. About this same time the Navajos and Paiutes began harassing the ranches and small settlements in southern Utah, probably in response to the rapid Mormon expansion into the region. Many of the smaller settlements, including Winsor, were abandoned as the pioneers grouped in forts and larger, better-protected towns until the threat was over. Finally, in 1871, the site was resettled and renamed Mt. Carmel, for a mountain in Israel. The Carmel Formation takes its name from this town.

54.7 ≈   The hills to the west are in the upper part of the Carmel Formation. Gray-white beds of gypsum mixed with siltstone were deposited in the shallow southern end of a narrow, sometimes restricted seaway that reached into the area from the north.

55.4 ≈   Yellow-brown rock in the roadcut on the west is limestone of the Co-op Creek Member of the Carmel. Scattered fossil bivalve shells can be found in limestone strewn along the shoulder, although fossils are more easily found in this unit just ahead, at Mt. Carmel Junction. Fossiliferous limestone suggests a normal shallow-marine environment. This is in contrast to overlying gypsiferous sediments, which indicate that this end of the sea periodically was barred from an influx of fresh sea water.

55.6 ≈   Mt. Carmel Junction consists of gas stations and motels that exist because of Zion National Park. Zion is reached by turning west on State Highway 9 at this junction. This route involves a 2-mile-long tunnel through the Navajo Sandstone and is not suitable for oversized vehicles.

The town of Kanab lies 17 miles south on U.S. 89. Kanab is the main headquarters for the Grand Staircase–Escalante National Monument. It is also the only town for some distance that is large enough to host a good-sized grocery store and a coffee shop/bookstore that carries outdoor gear and camping supplies.

15

Mt. Carmel Junction through Zion National Park to La Verkin via State Highway 9

State Highway 9 begins at Mt. Carmel Junction and extends west into Zion National Park, where, upon emerging from a mile-long tunnel, it rapidly drops into the bottom of the canyon. A 6-mile spur reaches up the North Fork of the Virgin River beneath the sheer walls and monolithic temples of the shaded canyon. This road ends at the Narrows trailhead. Highway 9 continues west out of the park to follow the course of the Virgin River through the towns of Springdale and Rockville. As the route approaches the western terminus of the Colorado Plateau, the road cuts through the Hurricane fault, dropping off its escarpment to enter the realm of the Basin and Range. The road log finishes at the town of La Verkin, which nestles precariously at the foot of the Hurricane fault, next to the Virgin River.

mile
0.0 ≈ Immediately upon our departure west from Mt. Carmel Junction we are confronted with a high exposure of the Co-op Creek Member of the Carmel Formation along the north side of the road. Several aspects of this well-stratified marine limestone make it especially interesting. First, throughout southern Utah we have yet to encounter marine limestone. Probably 90% of the marine limestones around the world, and throughout geologic time, have been born in a setting analogous to the modern Bahama Islands. An accurate portrayal would be deposition in a clear, warm, shallow sea with waves vigorously lapping at a shoreline of white sand. While in some instances the exact setting was undoubtedly more complex, that is the setting for the Co-op Creek Member.

Oolitic limestone can be found in the rock slabs that have fallen from the cliffs. This rock is composed of spherical grains of calcium carbonate called *ooids*. It closely resembles an ordinary sandstone, but it is packed with ooids rather than grains of quartz. These grains form in warm, agitated water saturated with dissolved calcium carbonate. As the fragments rolled back and forth along the shallow sea floor, calcium carbonate precipitated around them, forming a concentric accumulation no larger than a common sand grain. Ooid sands make up the gleaming white beaches of many Caribbean islands today.

Interbedded with oolitic limestone are fossiliferous limestones, mostly containing bivalve shells and crinoid stems. Crinoids, commonly called "sea lilies," were attached to the shallow sea floor. They had bulblike tops bristling with tentacles that waved in the currents. These heads were elevated off the sea floor on slender segmented stems that, upon death, broke into numerous round or star-shaped, buttonlike segments. Although the base and top of these organisms were rarely preserved, the stem fragments are a common component of marine limestones.

0.3 ≈   A small fault in the gully on the north side of the road drops red and white siltstone of the Crystal Creek Member on the left against the yellow-gray Co-op Creek Member on the right. This is one of a group of small, northwest-trending faults that fan out from the Sevier fault behind us.

1.0 ≈   Red-brown slopes lining the road are siltstone of the Crystal Creek Member. The thick white bed above it to the north is gypsum of the Paria River Member, also of the Carmel Formation.

3.2 ≈   Directly ahead in the distance is Clear Creek Mountain. Its slopes are composed of the Cretaceous sequence that was last seen in its entirety on the Kaiparowits Plateau.

6.7 ≈   As the road dips and swales across the new blacktop, the colorful Carmel Formation gives way to shades of gray and brown. We have entered a great landslide mass of Cretaceous rock that is inching its way toward Meadow Creek below. The mass originates high above, on the unstable flanks of Clear Creek Mountain. Mass movements are common in Cretaceous rocks throughout the Four Corners states because the abundant, organic-rich shale easily reverts to mud when saturated with water. The road continues to traverse this temporary landscape for the next few miles.

8.2 ≈   Looking southeast, we gaze down into the shadowed depths of the canyon cut by Meadow Creek as it feeds into the East Fork of the Virgin. The creek has sliced through the Carmel Formation and Temple Cap Sandstone and is floored by the Navajo Sandstone. Farther downstream the East Fork cleaves through an immense block of Navajo Sandstone to create the spectacular and remote Parunaweap Canyon, an unpopulated wilderness that surpasses even Zion Canyon in beauty.

10.6 ≈   We are following the drainage of Co-op Creek, the namesake for that member of the Carmel Formation. The Co-op Creek Member forms the low, tree-dotted hills that line both sides of the road.

10.8 ≈   The red-streaked, crossbedded cliff bands are the Temple Cap

Sandstone, a Jurassic eolian deposit that unconformably overlies the very similar Navajo Sandstone. The Temple Cap differs from the Navajo in that it contains thin red siltstone beds. It likely was more extensive in its original coverage, but pre-Navajo erosion reduced its preservation to this small southwestern pocket of the Colorado Plateau. The Temple Cap derives its name from its position above the Navajo Sandstone, which, as we will soon see, forms all the named "temples" that oversee Zion Canyon.

11.6 ≈ We find ourselves confronted with all the familiar characteristics of the Navajo Sandstone—huge, sweeping crossbeds, rounded domes, and soaring cliffs. The landscape again becomes a world of buff-colored slickrock, which will overshadow all vistas until we leave Zion.

12.2 ≈ We cross the boundary into Zion National Park.

12.8 ≈ While waiting to pay your dues at the Zion entrance station, look carefully at the surrounding walls. Differentiating between the Navajo and the overlying Temple Cap becomes a simple affair when both are exposed. The yellow Temple Cap tends to weather into a series of flat, tree-covered benches that recede from the top of the Navajo. Additionally, the crossbed sets in the Temple Cap are separated by thin red siltstone beds, giving the formation a distinct ledgy appearance. The red siltstone that typically lies between the two eolian formations is rarely exposed, but its presence can be inferred by the red iron oxide streaks that bleed down the uppermost part of the Navajo off the mesa top.

13.1 ≈ Checkerboard Mesa turnout: as we look south to Checkerboard Mesa, the origin of its name becomes evident (photo 17). The graceful curves of the giant crossbeds in the Navajo Sandstone are cut by the vertical furrows of joints. Crossbeds were forged by the incessant Jurassic winds as huge quantities of sand were molded into piles. The vertical joints are a later addition, generated by stress acting long after the ancient dunes had become brittle rock. The view before us is a classic one, often reproduced in textbooks to illustrate the unbelievable scale of crossbedding in the Navajo. For a clearer view of the joints, uncluttered by the sweeping crossbeds, we can turn to the opposite wall of the canyon behind us. Here the cleaved faces and sharp dihedrals of the jointed cliffs are more distinct. The flat top of Checkerboard Mesa is occupied by the Temple Cap Sandstone and the lower part of the Carmel Formation.

17.9 ≈ After tracing the deep meanders of the dry bed of Clear Creek that are cut into the Navajo, we approach the tunnel. This highway and the 1.1-mile-long tunnel were completed in 1930. Previously Zion Canyon could be reached only from the west, via the towns of Rockville and Springdale.

The Navajo Sandstone is one of the few rock units on the Colorado Plateau that could accommodate such a tunnel. The homogeneity of the

Photo 17. Checkerboard Mesa in Zion National Park is composed of the Navajo Sandstone. The checkerboard pattern is the result of horizontal stratification and crossbedding cut by subvertical joints.

sandstone coupled with its immense thickness made it a good choice for this undertaking. As we drive through the tunnel, the darkness is broken by a few tantalizing glimpses of red sandstone walls through windowlike openings. The road emerges in the Kayenta Formation and the upper slopes of Pine Creek canyon.

19.5 ≈   At the first switchback after exiting the tunnel, we gain a clear view of the incomparable Zion Canyon. The Navajo towers menacingly over the road here, its north-facing cliffs orange and black with lichens and moss. At road level is the crumbly sandstone and shale of the Kayenta Formation. As we look across Pine Creek to the opposite canyon wall, the thickness of sand that accumulated in this great Jurassic desert becomes stunningly apparent. The Navajo is more than 2,000 feet thick in Zion!

Below these sandstone monoliths are the red slopes and cliffs of the Kayenta. The cliff bands deserve a closer look. The continuous sandstone ribbons are actually composed of individual stacked and overlapping lenses, each outlining an individual river channel. Many of these discrete bodies are only fragments of sand-filled river channels. In most cases a large part of the original cross section was removed by erosion as subsequent rivers forced their channels into earlier deposits to establish their new course. Under careful scrutiny, however, individual channel forms can be recognized in the cliff bands. Slope-forming shale is the product of finer-grained floodplain deposition as the muddy Jurassic rivers sporadically overflowed their banks.

To the west the North Fork of the Virgin discharges from the narrow confines of Zion Canyon to flow south into a widening canyon. Above the canyon floor the great, overhanging Streaked Wall, named for the black water streaks that hang like tattered curtains, is capped by the Beehives, turrets carved into the top of the Navajo. To the right is the Sentinel, another fortification of massive Navajo Sandstone. At the foot of this towering mass is a large slump/landslide that was last active in the spring of 1995. As we will see time and again in the Zion area, gravity still has the upper hand. In fact, these narrow canyons have been dammed by landslides several times in the prehistoric past, forming temporary lakes in the canyon bottoms. As we continue down the switchbacks to the floor of Pine Creek, much of the road is cut into steep landslide debris that clings precariously to the slopes.

19.9 ≈   At the next switchback a huge alcove can be seen in the Navajo at the head of this box canyon. This arching alcove formed as a consequence of jointing. Large-scale parallel joint sets are an important factor in the evolution of the Zion landscape. The larger joints dominantly trend north-northwest and are easily recognized on topographic and geologic maps of the region. While Zion Canyon does not conform to their orientation, most of its side canyons, as well as those that feed Parunaweap Canyon, are large joints widened by the timeless passage of water. These fracture systems likely formed from compressional stress of the Laramide orogeny, which occurred from 65 to 50 Ma, or possibly may have been generated by more recent extensional stress associated with regional normal faults such as the Hurricane fault to the west. These stresses are acting on the west margin of the Colorado Plateau today.

22.0 ≈   Purple and white mudstone peeks out of the roadcut on the uphill side. This is our first encounter with the fluvial Moenave Formation, which replaces the eolian Wingate Sandstone that we have seen throughout the rest of southern Utah. The northwest-flowing Moenave river system marks the southwest margin of the Wingate erg and blocked its southward expansion. Interfingering between the Wingate and Moenave documents the turf battles between the two systems as the erg periodically pushed south, only to be thwarted by the big rivers that cut the great dune field back. Eventually the sand pile won the war: the Navajo represents its expansion over most of southwestern North America.

22.1 ≈　　　After we cross the Pine Creek bridge, the collection of Kayenta river channels is just overhead. Up close, the individual channel packages are separated by thin layers of red, purple, and white mudstone that form recesses. The basal parts of these sand-filled river channels contain pockmarks that at one time were pieces of the muddy floodplain that were ripped from their position by violent floodwaters and deposited in the bottom of the freshly excavated channel. They now form irregular cavities because the mud is more easily eroded than the sandstone that encases it.

Zion Canyon ≈

22.6 ≈　　　We turn north (right) into Zion Canyon to follow the North Fork of the Virgin upstream. The Kayenta lies across the river.

23.1 ≈　　　Driving up the canyon, we enter a zone of both prehistoric and recent landslides. Slopes on the east (right) side of the road consist of ancient landslide deposits.

23.5 ≈　　　On the west side, across the river, the steep, bare slope is the product of the most recent large-scale mass-wasting event in the region (photo 18). We are looking at a large landslide/slump complex that came back to life in April 1995. A wet spring caused this entire hillside to collapse into the canyon, destroying the road and damming the North Fork behind tons of debris. Water immediately began to pond behind the natural dam. The Park Service, fearing that a catastrophic breach of the dam would wipe out the valley floor downstream, quickly evacuated the campgrounds that lay downstream along the banks of the river. Fortunately the impounded water topped the debris pile and the incision was gradual, releasing the water slowly. Soon the river cut down to its original level and reestablished its normal gradient. The road was quickly rerouted and rebuilt around the toe of the slide. The new asphalt on the road and the retaining wall along the river mark the area most affected by the 1995 slide. Loose material on the steep slope across the river shows the activity is not yet over. Rocks and sand still occasionally bounce down the slope.

　　　Zion is, and always will be, a masterpiece in progress. While it could be readily accepted as a finished work, gravity is not satisfied. The great conveyor belt of the Virgin River constantly labors to transfer the rock mass it receives in the form of sediment. Gravity delivers the material, and the river whisks it away bit by bit to make room for more. In this way, over thousands and millions of years, this narrow canyon will slowly widen.

　　　If Zion Canyon is continually being widened, why is it so narrow, with

Photo 18. The 1995 slump/landslide that blocked the Virgin River in Zion for several days, backing water up into a temporary lake.

such steep walls? Although narrow canyons with vertical walls are a common feature of the Navajo Sandstone throughout the Colorado Plateau, the scale of Zion is exceptional. The Navajo is of the optimum consistency for deep incision, strong enough to support thousand-foot vertical walls, yet not so resistant as to escape being cut by the energetic rivers. But where do the rivers get the energy to cut down so rapidly? The answer may lie in the Basin and Range to the west and the interplay between tectonics (faulting) and erosion.

The energy level of any part of a river dictates whether it erodes, deposits, or does neither. A river's energy, in turn, depends on the gradient, the amount of vertical drop over a specified horizontal distance (e.g., feet per mile). Deficient-energy rivers have such a low gradient that they are unable to carry the sediment supplied by more energetic upstream reaches. In these cases, the sediment load is dropped, and deposition dominates. In contrast, excess-energy rivers are characterized by erosion, in many instances in the form of vigorous downcutting, which ultimately reduces the steep gradient. Excess-energy rivers are responsible for carving canyons. If unmolested, a river eventually should achieve a balanced state, where it has just enough energy to carry its sediment load but not so much that it erodes. This rarely occurs for any length of time, however, because the earth's surface is constantly being reshaped by tectonic activity.

Canyon incision in the Zion area likely was accelerated by regional tectonic activity. The Hurricane fault, which forms the western boundary of the Colorado Plateau, lies about 17 miles to the west (fig. 12.6). This fault has dropped the Basin and Range region on the west side as much as 7,000 feet relative to the uplifted Colorado Plateau on the east. Faulting increased the gradient and energy level of the rivers draining this margin of the Plateau, including the upper Virgin River and its tributaries. The most obvious effects are the deep canyons such as Zion that penetrate the Plateau edge.

In reality the Hurricane fault may be a minor factor in the carving of the Zion area. It is a larger but poorly defined event that generally is held responsible for most of the canyon-cutting throughout the Colorado Plateau. Sometime during the Middle or Late Tertiary, after deposition of the Claron Formation, the entire Colorado Plateau was uplifted. While the exact timing and mechanism for the uplift remain speculative, most geologists agree that the rapid erosion and downcutting that formed today's deep canyons, including the Grand Canyon, was triggered by this event.

24.3 ≈   From this pullout a short trail leads to a view of the high, multipeaked Court of the Patriarchs to the west. This cavernous box canyon of Navajo Sandstone is underlain by slopes of the Kayenta Formation. The base of these slopes, near river level, is marked by a cliff of crossbedded eolian sandstone. This sandstone bench represents a southward invasion of the mammoth dune field that blanketed the region to the north. But Kayenta rivers temporarily prevailed and regained their foothold.

Just below us along the river are relatively young laminated clay beds deposited in the quiet water of a lake or pond. It was the most chaotic of events, however, that created the lake that once stood here. About 4,000 years ago the collapse of the hillside at the foot of the Sentinel blocked the narrow canyon, forming a long narrow lake behind it. This slump was the precursor to the 1995 slump/landslide complex that we just passed through. As the water behind the obstruction deepened, clay settled to the lake floor. Where tributary streams fed into the lake, sandy deltas formed. Remnants of these lake deposits are evident along the river as we continue upcanyon.

The Narrows ≈

28.7 ≈ After passing beneath uncountable monoliths and shadowed alcoves, the Zion Canyon road ends at the Narrows trailhead. A one-mile walking trail continues upcanyon until the walls close in. This well-maintained trail ends when the canyon narrows considerably and wall-to-wall water forces the resolute hiker into the river. This deep, shaded chasm extends like this for 16 miles upstream, before escape from its depths is possible. In many places overhanging walls soar 1,000 feet above the canyon floor, which may be less than 40 feet wide at river level. One factor in the extreme downcutting is related to the river's extensive drainage network upstream. Because of this violent flash floods are a common occurrence: overnight backpack trips should be planned carefully if rain is even a remote possibility. Hikers are commonly trapped in the Narrows by flash floods, and many people have been killed in its depths when the narrow fissure suddenly became engorged with a wall of debris-laden water. Hikers are advised to use common sense and heed Park Service warnings.

The extreme constriction of Zion Canyon at the end of the walking trail marks the approximate point where the Kayenta Formation disappears into the subsurface to be succeeded by the Navajo Sandstone at river level. This resistant sandstone maintains its presence along the canyon floor for the next 16 miles. Over that distance the canyon passes from the base of the 2,000-foot-thick formation to its top. After that the Navajo drops beneath the surface, and the less resistant Temple Cap Sandstone replaces it at river level.

Along the walking trail and the lower canyon walls throughout Zion, lush hanging gardens glisten with water in the shaded alcoves and along the base of steep cliffs. These dense growths of fern, monkeyflower, and

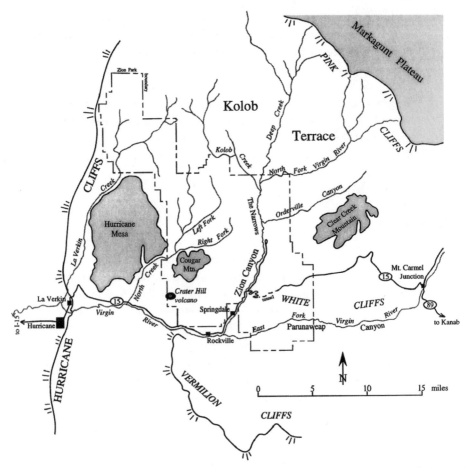

Figure 15.1. Map for the geologic road log from Mt. Carmel Junction through Zion National Park to the town of La Verkin. Prominent drainages, canyons, and physiographic features along the route are shown. The escarpment of the Hurricane Cliffs marks the trace of the Hurricane fault, a major normal fault that forms the boundary between the downdropped Basin and Range Province to the west and the uplifted Colorado Plateau to the east.

columbine are maintained by springs and seeps that issue from the porous rock. Seeps are concentrated along the contact between porous sandstone above and impermeable shale, siltstone, or limestone below.

This water begins its journey high on the plateau top hundreds of feet above. Rain, runoff, and snowmelt trickle into the porous surface of the Navajo Sandstone. Simple gravitational force pulls the water, drop by drop, downward through the rock, charting a tortuous path between the individual sand grains in the Navajo. The water descends slowly until it meets an impervious layer such as shale. At this point downward movement ceases, and the water is pushed laterally along the top of the shale, seeking relief from the pressure of still more water that falls from above. Relief is found in the form of an outlet along the deep canyon walls, where the shale/sandstone contact intersects the surface, and water again emerges to nourish the plants that could not normally survive the arid conditions.

The road log begins again at the intersection with Highway 15, where earlier we turned up the Zion Canyon road. The route continues west on 15, out of the park and along the Virgin River to La Verkin (fig. 15.1).

mile   We need to reset the odometer at the intersection between Highway
0.0 ≈   15 and Zion Canyon road. We turn right onto 15 to travel west.

0.1 ≈   After we cross the Virgin River bridge, the Moenave is partially exposed near road level, but mostly is mantled by detritus sloughed off the cliffs above.

0.9 ≈   As we continue downcanyon, the valley widens and the Moenave becomes well exposed along the lower slopes. The Moenave is mostly thin-bedded orange sandstone, purple shale, and white ledges of siltstone, all parts of a floodplain that lay adjacent to large northwest-flowing rivers.

1.4 ≈   Zion National Park exit: directly ahead, on the northwest side of the road, is a huge long-lived landslide mass that was last active in 1992. The hillside above hosted a small subdivision that was devastated as the hillside of loose rubble slumped and collapsed into the river valley, pushing onto the road that the steep, chaotic mass impinges on today (photo 19).

Movement of the slump mass took place on September 2, 1992, when several factors conspired to reactivate the loose slopes. The ground was saturated by rain, rendering the slippery shale of the underlying Chinle Formation unstable. But the real trigger came in the form of an earthquake that originated near the town of St. George, 28 miles to the west. The fault that generated the quake lay east of St. George but west of the Hurricane fault (fig. 12.6). This magnitude 5.8 earthquake, though some distance from Zion, apparently rattled the area enough that the weakened hillside failed, destroying three houses, two water tanks, and the infrastructure of the subdivision. Today only the abandoned skeletons of the houses and the impassable remnants of roads remain as a blunt reminder that gravity's business is unfinished.

1.9 ≈   Great mounds of rubble on the north mark the catastrophic activity of prehistoric landslides. Exposed among the debris piles are purple, gray, and brown humps of shale, the floodplain deposits of the upper Chinle

Photo 19. One of several abandoned houses above the town of Springdale destroyed by an earthquake-triggered landslide in 1992.

Formation. The north side of Springdale is built on hummocky landslide deposits, whereas the south side sits on the North Fork floodplain. This rapidly growing town appears to occupy a precarious position, although it has survived this location for well over a hundred years.

4.7 ≈   As Springdale and its motels are left behind, the road drops into a tree-lined canyon hemmed in by walls of tan sandstone and conglomerate. This is the Shinarump Sandstone, the basal member of the Chinle, which has not been seen since Capitol Reef. These fluvial deposits record the passage of large northwest-flowing rivers, probably larger than anything existing in western North America today. This great river system tapped sources as distant as modern-day Oklahoma and Texas.

Red thinly bedded sandstone and siltstone below the Shinarump mark the top of the Moenkopi Formation. These rocks represent the regular incoming and outgoing tidal currents along a monotonously flat coast. The line that separates the Moenkopi from the overlying Shinarump is an unconformity that marks the absence of any record for a span of about 20 million years, a period over which erosion won out over deposition. During this time dramatic changes occurred in the region. The climate shifted from extremely arid to seasonally wet; the depositional setting changed from the Moenkopi's shallow, featureless shoreline to raging rivers fringed by wetlands and thick forests in the Chinle.

5.7 ≈   We enter the town of Rockville (sign). Except for a few outsized houses, this community has managed to retain its small-town character. The canyon widens into a broad valley because the Moenkopi has been trimmed back to form a wider floodplain. Somewhere between Rockville and Springdale the East Fork of the Virgin River emerges from the gorge of Parunaweap Canyon to unite with the North Fork to form, simply, the Virgin River. This name is retained to its confluence with the Colorado River, which has been corralled into Lake Mead.

The waters of the Virgin River drainage are among the most endangered in North America. The rapidly growing retirement community of St. George covets these waters to supply the exploding population with its attendant grass-covered yards and golf courses. Luxuries such as these are efficiently killing the desert rivers throughout southwestern North America.

7.0 ≈   On the western outskirts of Rockville the valley grows still wider: the candy-striped Moenkopi that surrounds us has been carved into a maze of low hills and badlands. The Moenkopi thickens to more than 1,600 feet here, double the approximately 800 feet in Capitol Reef. This westward expansion reflects a deepening of the Early Triassic sea in this direction. The addition of several thick fossiliferous limestone units to the Moenkopi likewise suggests changing conditions. Carbonate-producing organisms require the continued presence of a shallow sea to colonize a region and thrive. They are particularly sensitive to fluctuations in water depth and the inflow of mud and sand that are responsible for the red and white tidal deposits seen over most of the Moenkopi's extent.

7.9 ≈   Black rubble ahead to the west is part of an extensive basaltic lava flow that is about 0.5 million years old. The flows are related to the small, well-formed volcano called Crater Hill that can be seen on the skyline of the mesa to the northwest. As fluid basalt poured from the earth, it spread across the mesa and inundated adjacent canyons, solidifying into resistant basalt dams. Coal Pits Wash was one such canyon, as evidenced by lake sediments in its upper reaches. After a few hundred years of impoundment behind this basalt plug, the water cut around it into the far weaker Mesozoic sedimentary rocks. The cliff-forming basalt caps the mesas along the north side of the road for the next 5 miles.

8.6 ≈   The wall on the north side of the road consists of relatively young conglomerate of the Pleistocene-age Parunaweap Conglomerate. Its age is constrained only as more than 0.5 Ma because the Crater Hill lava flows

Photo 20. Columnar joints in a young basalt flow above Highway 15 between Rockville and Hurricane.

cover it locally. These deposits recall a time when wetter conditions prevailed over the region.

9.3 ≈   As we traverse the base of the basalt-capped mesa, the view to the southwest makes the great thickness of the Moenkopi obvious: the gullied red and white vista stretches to the horizon.

9.9 ≈   The ragged Pine Valley Mountains loom to the west. This voluminous laccolith complex lies on the other side (west) of the Hurricane fault. An intrusive age of about 22 to 20 Ma makes this complex coeval with the Henry Mountains and the Abajo and La Sal Mountains to the east.

12.6 ≈   Basalt above the road displays well-developed columnar joints (photo 20). Although the exact genesis of these joint-bounded columns and their polygonal cross sections remains elusive, they result from cooling and contraction of the lava flow.

18.3 ≈   The Permian Kaibab Limestone can be seen to the south, at the top of the deeply incised canyon that is tributary to the Virgin River. The gray and tan limestone accumulated in the shallow sea that lay west of the rising Ancestral Rocky Mountains far to the east. This part of the sea was far enough west that it was unaffected by the detritus shed from these highlands. The Kaibab generally is rich with fossil brachiopods and crinoids. Its

Photo 21. Scablike fault surface of the Hurricane fault on Highway 15 as it drops into Hurricane. Vertical grooves were ground into the rock as the west side moved down relative to the east side.

proximity to the Hurricane fault, however, has fractured the brittle limestone to the point that such fine features are no longer recognizable.

19.8 ≈   As we drop off the Colorado Plateau, the roadcut in the shattered Kaibab Limestone demonstrates another effect of faulting. Movement along the Hurricane fault zone in which we now sit has dragged the Kaibab, causing it to become contorted by folds. Drag-folding along zones of fault movement is common.

20.2 ≈   The road down the steep hill follows the trace of the fault. Rocks that were caught in the fault are exposed on the east side of the road. Although the rocks are battered beyond recognition, this abuse is reflected in different ways over short distances. In some places the rock has been ground into a fine claylike material called *fault gouge*. In other places it has been broken into angular fragments of various sizes and recemented, forming *fault breccia*. The most striking fault-related features along this segment are in the solid rock along the road where vertical grooves have been etched into the fault surface (photo 21). This surface was polished and scraped as the bodies of rock on either side grated past each other. These grooves, known to geologists as *slickensides*, are excellent indicators of the relative motion between the two sides. In the case of the Hurricane fault the motion was vertical; the west side moved down relative to the east.

The Hurricane fault is a large recent fault that separates two major physiographic provinces of western North America, the Colorado Plateau and the Basin and Range to the west. The fault extends from the Grand Canyon in Arizona northward to the town of Parowan, Utah, a distance of more than 150 miles. The amount of vertical offset along the fault varies but generally increases from south to north. In the Grand Canyon area offset is estimated to be 200 feet, whereas a few miles north of where we stand, near the town of Toquerville, it reaches an incredible 6,800 feet! Most of this movement is believed to have occurred in the last 2 million years, and the fault is active today.

The recent population growth of southwest Utah so close to a large, active fault is alarming, as the chances of a large earthquake in the region are probably high. Because of this the Hurricane fault is currently the subject of intense scrutiny by geologists in an attempt to unravel its past history. It is hoped that more accurate predictions into its future activity can be made.

20.8 ≈   As we roll down the hill into the town of La Verkin, we leave the fault; the road descends a ramp of detritus that has collected at the foot of the Hurricane Cliffs, as the escarpment is known. Nested against the fault, La Verkin lies in an uncertain position. Like the rest of the region it is growing, and new houses can be seen abutting the fault scarp. This places these structures at "ground zero" for any earthquake generated by movement on this segment of the fault and directly in the path of rockfall that would most likely be triggered by even mild earth-shaking. We have observed firsthand how shattered the rock is that rests above town. It would not take much of a jolt to rattle it loose and allow gravity to take over.

Since dropping off the Hurricane Cliffs we have entered the Basin and Range. Although the rocks remain familiar until west of St. George, faults grow increasingly common, influencing the landscape.

This boundary we straddle has also been significant throughout much of geologic time. Many of the rocks we have passed through to the east were profoundly influenced by regional features that were situated here. If we had looked west from here during much of the Paleozoic Era, we would have seen the shallow ocean floor drop abruptly to great depths. Conversely, in the Late Jurassic and most of the Cretaceous Period we would have been confronted with mountains. The Sevier orogenic belt lay just west of us, its slow disintegration feeding sediment to the rivers and deltas represented by the rocks on the Kaiparowits Plateau and the Henry basin. Here also was the dividing line in the Early Tertiary when, about 65 million years ago, tectonic activity waned to the west, and the monoclines and modern Rocky Mountains of the Laramide orogeny were activated to the east. This history of division continues today as the Basin and Range evolves through large amounts of extension. To the west the crust is literally being pulled apart. Meanwhile, the Colorado Plateau sits serenely to the east, affected only slowly by grain by grain disintegration. Here, at the edge of the Basin and Range province, our journey ends.

References for Road Logs

Anderson, J. J., P. D. Rowley, R. J. Fleck, and A. E. M. Nairn. 1975. *Cenozoic geology of southwestern High Plateaus of Utah.* Special Paper 160. Boulder: Geological Society of America. 88 pp.

Best, M. G., E. H. McKee, and P. E. Damon. 1980. Space-time-composition patterns of Late Cenozoic mafic volcanism, southwestern Utah and adjoining areas. *American Journal of Science* 280:1035–50.

Billingsley, G. H., P. W. Huntoon, and W. J. Breed. 1987. *Geologic map of Capitol Reef National Park and vicinity, Emery, Garfield, Millard and Wayne Counties, Utah* (scale 1:62,500). Salt Lake City: Utah Geological and Mineral Survey.

Bowers, W. E. 1973. *Geologic map and coal resources of the Upper Valley quadrangle, Garfield County, Utah* (scale 1:24,000). Coal Investigations Map C-60. Washington, D.C.: U.S. Geological Survey.

———. 1975. *Geologic map and coal resources of the Henrieville quadrangle, Garfield and Kane Counties, Utah* (scale 1:24,000). Coal Investigations Map C-74. Washington, D.C.: U.S. Geological Survey.

———. 1990. *Geologic map of Bryce Canyon National Park and vicinity, southwestern Utah* (scale 1:24,000). Miscellaneous Investigations Map I-2108. Washington, D.C.: U.S. Geological Survey.

Davidson, E. S. 1967. *Geology of the Circle Cliffs area, Garfield and Kane Counties, Utah.* Bulletin 1229. Washington, D.C.: U.S. Geological Survey. 140 pp.

DeCourten, F. 1994. *Shadows of time: The geology of Bryce Canyon National Park.* Bryce Canyon: Bryce Canyon Natural History Association.

Doelling, H. H. 1975. *Geology and mineral resources of Garfield County, Utah* (scale 1:250,000). Bulletin 107. Salt Lake City: Utah Geological and Mineral Survey. 175 pp.

Doelling, H. H., and D. D. Fitzhugh. 1989. *The geology of Kane County, Utah: Geology, mineral resources, geologic hazards* (scale 1:100,000). Bulletin 124. Salt Lake City: Utah Geological and Mineral Survey. 192 pp.

Dutton, C. E. 1880. *Geology of the High Plateaus of Utah.* Washington, D.C.: Dept. of

Interior, U.S. Geographical and Geological Survey of the Rocky Mountain Region. 307 pp.

———. 1882. The physical geology of the Grand Canyon district. In *United States Geological Survey, 2nd annual report*, pp. 47–166. Washington, D.C.: Dept. of Interior.

Eaton, J. G., J. I. Kirkland, E. R. Gustason, J. D. Nations, K. J. Franczyk, T. A. Ryer, and D. A. Carr. 1987. Stratigraphy, correlation and tectonic setting of Late Cretaceous Rocks in the Kaiparowits and Black Mesa Basins. In G. H. Davis and E. M. VandenDolder, *Geologic diversity of Arizona and its margins: Excursions to choice areas*, pp. 113–25. Special Paper 5. Tucson: Arizona Bureau of Geology and Mineral Technology.

Flint, R. F., and C. S. Denny. 1958. *Quaternary geology of Boulder Mountain, Aquarius Plateau, Utah*. Bulletin 1061-D. Washington, D.C.: U.S. Geological Survey. 164 pp.

Geary, E. A. 1992. *The proper edge of the sky: The High Plateau country of Utah*. Salt Lake City: University of Utah Press. 282 pp.

Gilbert, G. K. 1877. *Report on the geology of the Henry Mountains*. Washington, D.C.: Dept. of Interior, U.S. Geographical and Geological Survey of the Rocky Mountain Region. 160 pp.

Gregory, H. E., ed. 1939. Diary of Almon Harris Thompson. *Utah Historical Society Quarterly* 7, nos. 1, 2, and 3. 138 pp.

———. 1949. Geologic and geographic reconnaissance of eastern Markagunt Plateau, Utah. *Geological Society of America Bulletin* 60:969–98.

———. 1950. *Geology and geography of the Zion National Park region, Utah and Arizona*. Professional Paper 220. Washington, D.C.: U.S. Geological Survey. 200 pp.

———. 1951. *The geology and geography of the Paunsaugunt region, Utah*. Professional Paper 226. Washington, D.C.: U.S. Geological Survey. 116 pp.

Gregory, H. E., and R. C. Moore. 1931. *The Kaiparowits region: A geographic and geologic reconnaissance of parts of Utah and Arizona*. Professional Paper 164. Washington, D.C.: U.S. Geological Survey. 161 pp.

Hackman, R. J., and D. G. Wyant. 1973. *Geology, structure, and uranium deposits of the Escalante quadrangle, Utah and Arizona* (scale 1:250,000). Miscellaneous Investigations Map I-744. Washington, D.C.: U.S. Geological Survey.

Hamilton, W. L. 1995. *The sculpturing of Zion with a road guide to the geology of Zion National Park*. Springdale: Zion Natural History Association. 132 pp.

Hereford, R. 1988. *Interim geologic map of the Cannonville quadrangle, Kane and Garfield Counties, Utah* (scale 1:24,000). Open-File Report 142. Salt Lake City: Utah Geological Survey. 16 pp.

Hunt, C. B., P. Averitt, and R. L. Miller. 1953. *The geology and geography of the Henry*

Mountains region, Utah. Professional Paper 228. Washington, D.C.: U.S. Geological Survey. 234 pp.

Lidke, D. J., and K. A. Sargent. 1983. *Geologic cross sections of the Kaiparowits coal-basin area, Utah* (scale 1:125,000). Miscellaneous Investigations Map I-1033-J. Washington, D.C.: U.S. Geological Survey.

Nelson, S. T., J. P. Davidson, and K. R. Sullivan. 1992. New age determinations of central Colorado Plateau laccoliths, Utah: Recognizing disturbed K-Ar systematics and re-evaluating tectonomagmatic relationships. *Geological Society of America Bulletin* 109:1547–60.

Nelson, S. T., and D. G. Tingey. 1997. Time-transgressive and extension-related basaltic volcanism in southwest Utah and vicinity. *Geological Society of America Bulletin* 109:1249–65.

Rusho, W. L. 1983. *Everett Ruess, a vagabond for beauty.* Salt Lake City: Peregrine Smith Books. 228 pp.

Sargent, K. A., and D. E. Hansen. 1982. *Bedrock geologic map of the Kaiparowits coal-basin area, Utah* (scale 1:125,000). Miscellaneous Investigations Map I-1033-I. Washington, D.C.: U.S. Geological Survey.

Smith, J. F., Jr., L. C. Huff, E. N. Hinrichs, and R. G. Luedke. 1963. *Geology of the Capitol Reef area, Wayne and Garfield Counties, Utah.* Professional Paper 363. Washington, D.C.: U.S. Geological Survey. 102 pp.

Stephens, E. V. 1973. *Geologic map and coal resources of the Wide Hollow Reservoir quadrangle, Garfield County, Utah* (scale 1:24,000). Coal Investigations Map C-55. Washington, D.C.: U.S. Geological Survey.

Van Cott, J. W. 1990. *Utah place names.* Salt Lake City: University of Utah Press. 453 pp.

Williams, V. S., G. W. Weir, and L. S. Beard. 1990. *Geologic map of the Escalante quadrangle, Garfield County, Utah* (scale 1:24,000). Map 116. Salt Lake City: Utah Geological and Mineral Survey. 6 pp.

Woodbury, A. M. 1950. The history of southern Utah and its national parks. Reprinted from *Utah Historical Quarterly* 1, nos. 3 and 4 (1944):111–224.

Zeller, H. D. 1973a. *Geologic map and coal and oil resources of the Canaan Creek quadrangle, Garfield County, Utah* (scale 1:24,000). Coal Investigations Map C-57. Washington, D.C.: U.S. Geological Survey.

————. 1973b. *Geologic map and coal resources of the Carcass Canyon quadrangle, Garfield and Kane Counties, Utah* (scale 1:24,000). Coal Investigations Map C-56. Washington, D.C.: U.S. Geological Survey.

Acknowledgments

I would like to give my heartfelt thanks to my parents Robert and Linda Fillmore for letting me figure it out on my own and allowing me to believe I was doing the right thing; people with whom I have shared time in this part of the world, Doug Scott, Dan and Steve Cutler, John Fletcher, Kathy Brown, Dave Huffman, Matt Kaplinski, Joe Hazel, Eric Martin, Mike Doe, Mark Martin, Drew Coleman, Chubs, and the geology students of Western State College; and my teachers and good friends, Bruce Bartleson, Allen Stork, Tom Prather, Larry Middleton, Ron Blakey, and Doug Walker.

Finally, I owe a huge debt to all the geologists who have worked in this fantastic region; much of this book comes from their work.

Index